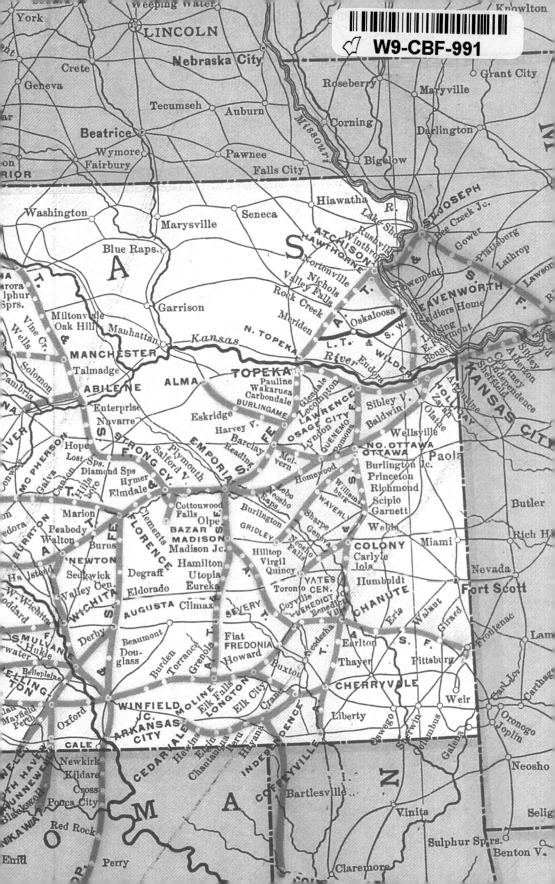

AVID

READER

PRESS

FROM THE
RIVER
TO THE
SEA

THE UNTOLD STORY
OF THE RAILROAD WAR
THAT MADE THE WEST

JOHN SEDGWICK

AVID READER PRESS

New York   London   Toronto   Sydney   New Delhi

AVID READER PRESS
An Imprint of Simon & Schuster, Inc.
1230 Avenue of the Americas
New York, NY 10020

First Avid Reader Press hardcover edition June 2021

AVID READER PRESS and colophon are trademarks
of Simon & Schuster, Inc.

For information about special discounts for bulk purchases,
please contact Simon & Schuster Special Sales at 1-866-506-1949
or business@simonandschuster.com.

The Simon & Schuster Speakers Bureau can bring authors to your
live event. For more information or to book an event contact
the Simon & Schuster Speakers Bureau at 1-866-248-3049
or visit our website at www.simonspeakers.com.

Manufactured in the United States of America

1 3 5 7 9 10 8 6 4 2

Library of Congress Cataloging-in-Publication Data is available.

ISBN 978-1-9821-0428-3
ISBN 978-1-9821-0430-6 (ebook)

For Patrick McGrath

Contents

PART THREE: LOS ANGELES

Each was what the other had not chosen to be, the cast-off self, what he thought he hated but perhaps in reality loved.

—Patricia Highsmith, *Strangers on a Train*

A Tale of Two Cities

L.A. THE ONLY CITY IN THE WORLD THAT GOES JUST BY its initials, like the self-assured global celebrity it is. Unlike Miami with its beaches, New York with its skyline, or Houston with its oil, Los Angeles is a fantasy of a city whose identity somehow floats free of mundane physical characteristics. All, that is, except for the sunshine radiating down from impossibly blue skies and the palm trees that rise up in greeting.

Unlike virtually everywhere else in America, to say nothing of America itself, L.A. has no founding myth to define it. No pilgrims, no explorers, no pioneers. While most people have the vague idea that the city dates back to Spanish times, the details are lost in the glitter, replaced by the gauzy notion that it somehow created itself as a product of its movie business.

It's hard to account for it otherwise. Although L.A. lies by the sea, it did not begin life as a port. Nor was it birthed by the river that runs through it from the San Gabriel Mountains or a natural resource like the gold that brought prospectors surging into San Francisco. (Oil wasn't found until L.A. was well established, which is why a pumpjack might be cranking away in a McDonald's parking lot.)

No, the city in fact owes its origin to something so foreign to its self-conception that it represents a violation of its existential code. It was started by a railroad. Los Angeles is a railroad town. Startling as that might sound, on reflection it should not be quite so surprising, since railroads gave rise to countless cities in the West (and plenty in the East, too). While San Francisco, up the coast, was not built by a railroad, it was certainly built *up* by one when the first transcontinental arrived there in 1869. Numerous other western cities were created almost entirely by railroads—Denver, Reno, Dallas, Houston, Seattle, Tacoma, to name just a few.

Curiously, Los Angeles was not the result of the first railroad that came to town nearly so much as the second. Its arrival set off a furious competition between the two in the spring of 1887 that dropped the price of a $125 ticket from Chicago to just one single solitary dollar. The news set off a stampede into Los Angeles. Just in the first three years of the frenzy, it went from a sun-splashed Spanish pueblo of thirty thousand to a bustling city of a hundred and fifty thousand, a fivefold expansion that marks the most explosive growth of any city in the history of the United States. That growth curve has rarely flattened since.

Over a thousand miles to the east, Colorado Springs lies just south of Denver on the edge of the Rockies, a mile up in the crystalline mountain air. A rather sedate, if not sleepy, college town in the shadow of Pikes Peak, a jagged-topped "fourteener" that looms over everything, Colorado Springs was also created by a railroad. Founded in 1871, it was intended to be a mountain retreat in the Alpine manner, a place of healthy air and cultural refinement for high-end refugees brought in by train from the smoggy East. Small, out-of-the-way, closed-in, Colorado Springs seems to exist in a separate universe or on a separate plane of meaning from L.A. But there is a connection between them all the same.

The train that made the modern Los Angeles started in Colorado Springs. Not literally—the town never had an L.A. Express—but figuratively, riding the tracks of history, which often run by puzzling, circuitous routes from the past into the present. While the course of progress is often thought to be the result of economic, social, technological, and environmental forces beyond anyone's control, that was not at all true of the development of the railroads. In the robust industrial age, they were all run by powerful, strong-minded men who bent their industry, and a good deal of the country, to their will. They set the course, chose the route, and built up the cities and towns their tracks reached. In this, Colorado Springs and Los Angeles were no exception.

The fates of these two distant cities, one as big as the other is small, were linked because the railroad men behind them were linked. More than linked, in fact. Bound like a pair of conjoined twins, two bodies somehow sharing a single mind, burning as one with the identical, all-consuming determination to go west. It was freakish, but undeniable: these two wildly different men became almost indistinguishable once they focused on the same objective and did so in the full realization that only one of them could attain it. It made quite a ball of fire, this frenzied competition, a blind, stupid, and utterly destructive jealous rage. A sun all of their own making that drew all eyes to it—even as the real one rose up overhead, day after day, and silently crossed the sky to the far horizon, as if to remind these two railroad men what they were fighting for: the chance to develop and define the modern West as no one else could.

A Very Personal War

WILLIAM BARSTOW STRONG AND GENERAL WILLIAM Jackson Palmer met three times over the course of their decade-long fight to run tracks from Colorado to the sea, but the visits did nothing to warm the two men to each other. The first was at the General's castle in Colorado Springs, a handsome wooden fortress of high ceilings and stunning views that was inspired by an ancient version in Scotland. He put it up amid some ruddy, up-thrusted sandstone formations in a near-sacred spot called the Garden of the Gods. Colorado Springs was that kind of place, and the General very much that kind of man. The meeting was in early November 1877. The General, who had been raised in Philadelphia, had built Glen Eyrie five years before at the age of thirty-six to lure Queen Mellen, a wild-haired nineteen-year-old beauty from Flushing, New York, to come live with him as his wife in western splendor. To him, the jagged, snow-capped Rockies were nothing less than an earthly paradise. "A sight burst upon me which was worthy of God's own day," he wrote Queen after seeing them for the first time. "The Range, all covered with snow, arose, pure and grand, from the brown plains. As I looked I thought, 'Could one live in constant view of these grand mountains without being elevated by them into a lofty plane of thought and purpose?'" That vision lit him up, but his Queen was at heart a city girl, and, as his letter demonstrated, she needed some convincing.

General Palmer had been a certified Civil War hero before turning to railroading. He was an extraordinarily handsome man with swept-back hair, a well-turned mustache, sky-blue eyes, a proper, military bearing, and the air of breezy self-confidence that can arise from such qualities. Turned out for photographs in a flowing jacket, taut vest,

General William Jackson Palmer.

wing collar, and necktie, he had to have been the best-dressed man in Colorado. If other subjects in that early photographic era often seem a bit startled by the pop of the photographer's flashbulb, the General maintains his poise, unperturbed.

Palmer signed up for the Civil War after considerable soul-searching. Few take easily to battle, but in his case, the anguish was personalized. Raised a thee-thouing Quaker, he agonized over whether his commitment to ending the horror of slavery outweighed the strict pacifist obligations of his religion. Ultimately, while he conceded war was "inconsistent" with "the example of Jesus Christ" as his parents had stressed, he decided "the inner light made it very plain" he still needed to join up. Once he did, though, he was overjoyed to plunge "into the maelstrom," as he put it, in the cavalry of the Pennsylvania Fifteenth Volunteers, which he considered a better class of men. After Antietam, he volunteered to serve as a spy behind enemy lines to see if General Lee planned a retreat—only to be caught when a mindless companion gave him away. Sentenced to be hanged by order of Lee himself, Palmer was thrown into a hellhole Confederate prison instead, a stroke of luck that saved his life. He tried to escape by sawing through the floorboards under his bed with a serrated jackknife, only to discover that the floor stood on pilings over bare ground, with sentries all around. After months of imprisonment, he finally cadged his freedom, returned to the war, and won the rank of brigadier general for his bravery at the otherwise disastrous battle of Chickamauga in the western theater of Tennessee. After Appomattox, he completed his service by tracking down the fugitive Confederate president, Jefferson Davis, in Georgia.

Soldier, spy, hero. Palmer's war years made for quite a résumé, and now he hoped to make something of them in the railroad business.

Palmer had started railroading at seventeen when he did some surveying for a small line seeking to push a train through the Alleghenies. By the time the war broke out, he'd risen to become the personal assistant of the Pennsylvania Railroad's president, J. Edgar Thomson.

When Palmer returned from the war a hero, Thomson appointed him to an executive position at the Kansas Pacific, an offshoot of the Pennsylvania that was running west from Kansas City.

Besides his lengthy railroad experience, it was his warrior side, plus some army connections, that the General relied on most to create his Denver & Rio Grande Railway Company in 1871. Inwardly, however, he had some gentler characteristics that he revealed almost exclusively to Queenie. To her he was tender, dreamy, and not a little poetic in his aspirations to build his railroad, to take the West, and to win her hand. By 1877 the construction of the General's railroad was well underway, and he had built not just his Colorado Springs castle for Queen, but Colorado Springs itself.

It was fitting that he would place his town below Pikes Peak, the most forbidding mountain in the Rockies, about a hundred miles down

Palmer at thirteen, with his parents,
John and Matilda.

from Denver. Palmer was always stirred by the sublime, which to him had some of the majesty of death itself. With that jagged, ice-topped peak as the looming backdrop, Palmer built Colorado Springs to his exacting specification, personally laying out the streets, positioning the parks, and even deciding on the citizenry to make sure they would be the perfect sort for his wife. He made Colorado Springs less a town than a social club of which he alone determined membership, restricting it to wealthy, cultured, well-connected, and urbane easterners and Europeans who would appeal to his citified wife. And he banned alcohol to distinguish his settlement from the boozy hell-on-wheels towns that were popping up elsewhere in the West, bringing nothing but embarrassment with them. The General made the town the first stop south on his new Rio Grande railway, and, although the line would ultimately extend more than a thousand miles more as the General chased his dreams, it was always the place of his heart, for he saw it as the place to win the heart of his Queen.

The General's first meeting with William Barstow Strong at Glen Eyrie came shortly after Strong joined the Atchison, Topeka and Santa Fe Railroad as its general manager, charged with taking that small but ambitious line west. The Santa Fe had only recently crossed from its native Kansas into the General's home state of Colorado, which was why Palmer was so keen to meet him. He was eager to size up the interloper and see what he could do about keeping him from causing any trouble.

At that point, the Union Pacific and Central Pacific had already famously joined at Utah's Promontory Point in 1869 to create the nation's first transcontinental. That line ran well to the north, on Chicago's latitude across Wyoming to San Francisco, deliberately skirting the highest of the Rockies. The Northern Pacific was slated to run west even farther north, closer to the Canadian border, to Seattle; it had been started in 1864, but had tumbled into bankruptcy and now lay unfinished. The Southern Pacific, built by the backers of the Central Pacific, was snaking south from San Francisco in search of a southern route east through Yuma, at the bottom of the Arizona Territory. To these other railroad men, the Rockies marked a frightening, no-go zone, but Strong and the General had seen an opportunity, and dared to seize it.

The Santa Fe Railroad was owned by its investors, an assortment of interbred Bostonians who were known collectively as the "Boston Crowd." They established the corporate headquarters on Devonshire Street in downtown Boston, nowhere near the West they were seeking to claim. For the Santa Fe, unlike Palmer's clubby Rio Grande, which he largely owned and solely operated, was that new thing largely of the railroads' creation, a modern corporation. In that hypercapitalist era the idea was not, in fact, to build a railroad so much as to create a financial vehicle to make money off a railroad, while leaving its actual management to trained professionals like Strong, who had little financial stake in the enterprise. Still clinging to the old ways, Palmer saw a corporation like the Santa Fe as a Frankenstein's monster, partly human in that it could hold property, sue, and be sued, and partly a machine in that it was just a mechanism to bring wealth to its investors, while freeing them from the full legal, financial, and moral consequences of proper ownership. It was stuffed with people but had no personality.

That was never Palmer's way. While he relied on financial backers in Europe and the East, he viewed his railroad company as his "little family," with himself at the head, the proud papa. It was a point of particular satisfaction for him. As the third of four children, Palmer had been the fragile one, bedridden for months at a time with a variety of severe, but unspecified, complaints. Now, let other railroads like the Santa Fe grow into vast, disembodied, highly capitalized corporations; the General was determined to keep the Rio Grande a sole proprietorship, guided solely by his own desire to capture the soul-stirring beauty of the Rockies. He brought others into it, of course, not as investors, but as *friends* who felt the Rio Grande was, as he said, "their own road, and not some soulless stranger corporation."

Reflecting the intimacy of the small Palmer's line was known as the "Baby Road" for its almost unique reliance on three-foot-wide "narrow gauge" tracks that Palmer considered more efficient and economical when twisting about the mountains, as opposed to the Santa Fe's "standard gauge" width of four feet, eight and a half inches. Palmer was convinced that small was not just beautiful, but in the uncertain world of the West it was practical and safe. He held firmly to that belief even as more railroads adopted the standard gauge. The General often paid little heed to what others did. If other trains did not adapt to his preferences, that was their mistake.

Strong had little use for small, and even less for feelings. Not in business, anyway. If the Santa Fe was not yet big, he knew it would get big, and then bigger still, and it fairly exulted in the wider gauge. It would eat cash and grow. Barreling out of Kansas, it was built for the long haul, with powerful locomotives up front and plenty of cars behind. Strong had no use for the light and nimble. He trusted in sheer might. Bigger was always better. No freight car could ever become too bulky, no passenger cars would ever be too numerous to fill, no locomotives would ever become too heavy for the rails they rode on, and, more broadly, no highly capitalized railroad corporation would ever become too unmanageable to get west. Of all this, Strong was utterly convinced. With its size and weight, the Santa Fe was designed to burst through obstacles, not twist around them.

From these crucial differences between the General's Rio Grande and Strong's Santa Fe one rose above all the others: their geographical orientations, a full ninety degrees apart. Like just about every other railroad with transcontinental ambitions, the Santa Fe planned to get west by going west. Coming from Kansas, the Santa Fe was on a line to California that a bullet might take. The General, on the other hand, planned to get west by going *south*. His Denver & Rio Grande would hold true to its name and run down to Mexico like the Rio Grande

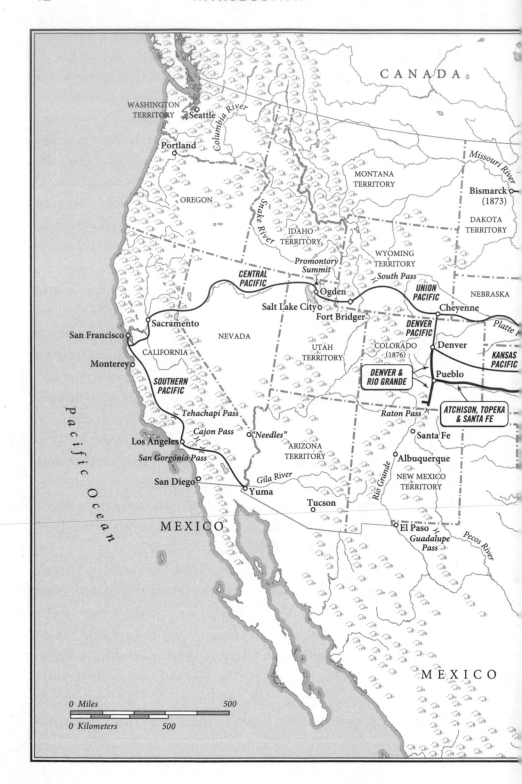

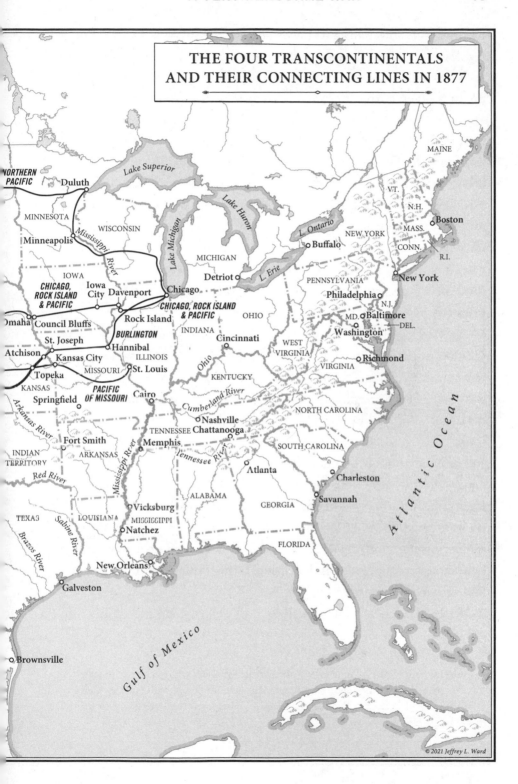

THE FOUR TRANSCONTINENTALS AND THEIR CONNECTING LINES IN 1877

NORTHERN PACIFIC

Duluth

Lake Superior

MAINE

VT.

N.H.

MINNESOTA

WISCONSIN

Lake Huron

L. Ontario

NEW YORK

MASS.

Boston

Minneapolis

Mississippi River

Lake Michigan

MICHIGAN

Buffalo

CONN.

R.I.

IOWA

Detriot

L. Erie

PENNSYLVANIA

New York

CHICAGO, ROCK ISLAND & PACIFIC

Iowa City

Davenport

Chicago

Philadelphia

N.J.

Omaha

Council Bluffs

Rock Island

CHICAGO, ROCK ISLAND & PACIFIC

OHIO

MD.

Baltimore

DEL.

BURLINGTON

INDIANA

Washington

St. Joseph

Hannibal

Cincinnati

WEST VIRGINIA

Atchison

Kansas City

ILLINOIS

St. Louis

Ohio

VIRGINIA

Richmond

Topeka

MISSOURI

KENTUCKY

KANSAS

PACIFIC OF MISSOURI

Cairo

Cumberland River

NORTH CAROLINA

Springfield

Arkansas River

Nashville

TENNESSEE

Chattanooga

Fort Smith

Memphis

Tennessee River

SOUTH CAROLINA

INDIAN TERRITORY

ARKANSAS

Mississippi River

Atlanta

Charleston

Red River

ALABAMA

GEORGIA

Savannah

TEXAS

Sabine River

LOUISIANA

Vicksburg

MISSISSIPPI

Natchez

FLORIDA

Brazos River

New Orleans

Galveston

Atlantic Ocean

Brownsville

Gulf of Mexico

© 2021 Jeffrey L. Ward

River, which flowed from its headwaters in the snowmelt of the Rockies down along the Texas border to empty into the Gulf of Mexico at Brownsville. More outlandishly, Palmer dreamed that his railroad might go even farther, pushing through Mexico's capital city clear to a Pacific port on the country's Gulf of California. He'd connect his line to ships bound from there not just to California, but to Australia and Asia as well. The idea wasn't entirely preposterous. If shippers who'd otherwise dock in San Francisco Bay could get past the idea of Mexico as a foreign country, the route would save them well over a thousand miles. By going south, Palmer would also have all the connecting business between any further east-west lines to himself, and—this was always important to him—by going south he could keep competitors like the Santa Fe guessing about his intentions, since he could veer west from any point along the route

In March of 1876, the two lines, the Santa Fe coming west, the Rio Grande running south, had crossed at Pueblo, about forty-five miles down from Colorado Springs. That intersection, long anticipated by the Santa Fe, gave Palmer a jolt. It seems never to have occurred to him that another line would ever encroach on his territory, and for some time he was flummoxed about how to respond. Some of his distress stemmed from the natural anxieties of railroad men everywhere, caught in the ferocious, big-money Darwinian struggle for survival in uncertain territory. But the stress was far worse in the West with its uncharted, mostly inhospitable terrain of barren desert and steep mountainside. Not to mention the nasty and perilous surprises like venomous snakes, hostile Indians, gun-toting outlaws, drunks, plagues, and horrendous weather.*

In 1877, General Palmer still had all of Colorado pretty much to

---

* I use the word "Indians" to refer to America's indigenous people because that was the term in general use at the time, and the one they used to describe themselves. Being of relatively modern coinage, "Native Americans" would have been unknown to them. I mean no disrespect.

himself, so it didn't much matter what others did. But, of course, as other lines like the Santa Fe encroached upon his, he ran the risk of being isolated, his network of trains an island, as his adherence to narrow gauge would require awkward and time-consuming transfers if the Rio Grande were to pass its freight and passengers to the standard gauge of other lines, or other lines passed theirs to his. To a railroad man, the greatest terror of all was another train coming into territory he'd thought was his alone. It was frightening enough to see a rival's tracks come closer, set down by teams of fevered workmen, rail by rail. It was far more disturbing to find tracks that had somehow materialized overnight, which could happen at any time once a competitor was in range. Vast as the West might be, a competitor made it all too tight.

And now, here a competitor was.

►─┤─◆─○─◆─┤─◄

As soon as Strong stepped inside the heavy, wide door of Glen Eyrie, he was greeted by Palmer in his front hall. Immediately, they both must have felt a rivalry stir within them, Strong's meatier hand enclosing the General's more delicate one for an introductory shake. That must have been discomfiting for the General. Although he was only a year younger, forty to Palmer's forty-one, Strong was supposed to be the obsequious junior man. At that point, Palmer had owned his Rio Grande for six years. How long had Strong had been on board the Santa Fe—a week?

The General must have felt a chill, too. Strong was, to him, an alien figure, odd to the point of being inscrutable. Palmer always drew his closest associates from the old soldiers of the Fifteenth Pennsylvania, but Strong had not served, and therefore offered none of that old-soldier camaraderie. Nor was he a man of the Rockies. More startling still, in an era when men were defined by their whiskers, he wore a long, wavy Old Testament beard that ran down from his chin to his chest like an inverted flame, and it shook in a grandfatherly way when he spoke. It somehow made him all beard. All, that is, except for the bracing determination in Strong's eyes under his stern brow. His face

William Barstow Strong.

was not tanned or weather-beaten from the rugged, outdoor life, but instead bore the sallow pallor of the career office dweller. This was, in fact, Strong's very first time west of the Missouri. A midwesterner by upbringing, he worked out of Santa Fe's tiny, brick-fronted operational headquarters in downtown Topeka, Kansas.

Nonetheless, Palmer may have sensed uneasily that Strong saw *him* as the strange one. The General was likely wearing jodhpurs, tweed, and riding boots, his usual getup at home, and sported a clean-shaven face, except for a neatly trimmed mustache. (The workaday Strong routinely stuck to a business suit and tie shoes.) There was the matter of the General's accent, too. Having spent some time in England, he'd developed an English accent to go along with his proper English spellings.

While many admiring photographs of the General capture him from head to toe, the precious few pictures of Strong are only of his bulky upper body, suggesting he was a far bigger man than the General, who was just five foot eight. But that was just a part of it.

For a titan of the railroad industry, the details of Strong's life are oddly scant even now. We know only that he was born in 1837 in Brownington, Vermont, up by the Canadian border, and he was descended from Puritans who'd come to America in 1630. Although Strong's father had been high sheriff of the county, he and his wife took the family to Beloit, Wisconsin, a town founded on the New England model by emigrants from Vermont in 1836. There, the couple established a temperance hotel—and likely encouraged their sons' higher aspirations. For the oldest became the mayor of Beloit and the second the president of Minnesota's Carleton College. The third was William.

He started in with the railroads at fifteen, taking a position as a railway agent and telegrapher for the Milwaukee & Mississippi at its Beloit station. He soon moved from there to the M & M's stop in Milton, then to two more Wisconsin towns. Although one brother served in the Civil War, William stayed with the railroads. He caught on at ever higher positions with two other, larger railroad lines coiling about the state, culminating in 1876 with an appointment as general superintendent of the potent Chicago, Burlington and Quincy, another Boston-based railroad that was eager for better things. Just one year later, the Santa Fe snapped him up to be its general manager in Topeka, in charge of directing the western expansion. By then Strong had married a Beloiter, Abbie Jane Moore, and had three children with her. His family did not come to live with him but settled instead into a modest house in Chicago while Strong toiled for the railroad down in Kansas.*

Of Strong's private life, the inner man, little written evidence survives. Just about all of it resides in a scant "letter press" collection of blurry carbon copies of office correspondence he marked "personal," from the early years of his presidency. Nonetheless, those letters reveal a good deal about the force of his personality, as all the writing is laid down in the brisk hand of a man with far too much to do; he is clear-eyed in his objectives and invariably comes right to the point. Crisp in their succinctness, his letters stand in sharp contrast to Palmer's, which could be windy. A favorite term of approval for Strong, aptly enough, is "strong," which for him conveyed conviction. "Of course we should make ourselves stronger and stronger every day," he wrote to a friend in the business office. "Do not loan any more money at present. Collect all you can—And let us be very strong in actual money at hand." But, more than the General, he had a marked preference for firm and decisive action. No dithering for him. "Be ready to move," he told another associate. "And 'in full blast.'" And his principles come through, too. "What I desire in his case is not what will be best for him," he writes of one subordinate, "but what will be best for the company and the question is wholly in your hands for a solution." For all his executive command, Strong is not without fellow feeling. A brotherly concern for favored associates runs all through these few letters. He shows, for example, a

---

* All but the last of these details come from a three-page entry out of the thousand-plus pages of the fat, two-volume history of Rock County, Wisconsin, published in 1908, which offers the most comprehensive account of Strong's life in existence, even though it skips over his years with the Santa Fe almost entirely, and offers an apt summary of the man: "His knowledge of human nature and trained judgment enabled him to pick the right men for the right places, and his genial nature secured from all subordinates their personal devotion and very best service." Recognizing Strong's position as the county's most famous resident, the book takes Strong's austere, full-bearded image for its frontispiece. There is little else. Of the 1,700 cubic feet of the complete records of the Santa Fe Railroad in the Kansas Historical Archives in Topeka, Strong's personal file takes up about a quarter of one single cubic foot. Originally, there had been two boxes; the other has gone missing.

touching sympathy for a colleague who'd been experiencing "the blues." More remarkably, he extended that same consideration to himself in a fashion that is stunning for a titan in an industry that favored bluster over candor. In one dire interval, he confided in a colleague to a debilitating despair when it appeared his railroad might be doomed.

Broadly, Strong spoke in actions, which were communicative enough, and recorded in the network of the train lines he built. His answer to every business question was to lay down track, and then to lay on more. "A railroad to be successful must also be a progressive institution," he wrote. And by progressive he did not mean politically. It must go forward. *Must!* "It cannot stand still, if it would. If it fails to advance, it must inevitably go backward and lose ground already occupied." Grow or Die—this was the maxim Strong lived by, pushing his railroad ever deeper into the uncharted West, no matter what obstacles or hazards. He intended to push his railroad to the Pacific if he had to put a shoulder to it himself.

Now, as he stood in the General's front hall, Strong made clear this was not a social call. He had a proposition for him. He'd like to lease the General's little railroad, thirty percent of it, to be exact. Palmer's jaw surely dropped at that. He could not have been more dumbstruck if Strong had asked to borrow Queenie for the weekend. *What on earth?* Palmer was too much a gentleman to respond with the contempt that in his mind such a question deserved. Instead, he demurred, likely between clenched teeth, his heart pounding in fury. It was an absurd idea, unbearable, not to be considered. When Palmer later reported Strong's outrageous offer to his general manager, he let him know that it was to go no further. "Keep this carefully and confidential, or destroy it," Palmer told him. As far as Palmer was concerned, his meeting with Strong was over.

>-+-◦-◦-+-<

When two trains drew into close proximity, the question in railroading was whether to cooperate or to compete. Cooperate meant sharing tracks and pooling revenues; compete meant engaging in rate wars and legal strife. In Palmer's hall that afternoon, the answer was clear. Compete everywhere over everything. Even though the West could not have been more wide open in those years, Palmer and Strong ended up fighting wherever they went. While they could have avoided each other, they moved closer together time after time, with devastating results. It was almost as if they had telepathic powers, a feeling for how each other was going to act—and then tried to do it first themselves. The two lines could never break free of each other, and neither could the two men. They lived in each other's shadow, for each the clearest proof of a route's value was that his rival wanted it. For two years they battled to take all of Mexico, but they converged most crucially at two narrow passes through the Rockies where there was room for just one set of tracks. Still, the conflict ran deeper than a fight for any individual piece of territory. Both men set their sights on that distant, glittering shore of the Pacific to which all Americans' eyes were increasingly drawn in that expansionist age.

Because the vast terrain of the great Southwest was so open, there was an empire to build along the way. In 1877, the General had planted his flag at Colorado Springs and was working his way south from there, building towns as he went. He'd gotten as far as El Moro, a coal-based railroad town of his creation at the very bottom of the state. Since the Santa Fe had only recently reached Pueblo, Strong had scarcely gotten started.

Still, both men were intent on raising civilization from the barren sea of wildness that was the West. A railroad would string together the few settlements and fewer true cities to form archipelagos that rose out of the primordial waters, and they would, in turn, rise and spread into wide peninsulas of lively business that ultimately merged into a whole new landmass of cities and towns clear to the far coast, all of it studded with handsome buildings along proper streets, joyous with parks, abuzz with people, and throbbing with the commerce that was, in the end, the railroads' ultimate product.

This would be the new West these two railroad men sought to build out of the nothingness of the old. But only one could win. And which of them would it be? Would it be Strong or the General? The manager or the proud papa? The corporation or the sole proprietor? The machine or the man? Would it be mass-produced or handcrafted? Driven by greed or inspiration? There was much in play, and all of it fed a conflict that spun out from the personal to the professional to the existential and eventually led to an armed confrontation that resulted in the longest, most expensive, and most destructive railroad war in American history.

An early Rio Grande train passing through the Colorado Rockies.

PART ONE

THE RATON PASS

CHAPTER 1

>-+<>-0-<>+<>+<>-0-<>+<

On the Train to El Moro

SHORTLY BEFORE SEVEN O'CLOCK ON THE BITTER, snowy evening of February 26, 1878, two men, both traveling lightly but well bundled against the cold, boarded a Rio Grande train in Pueblo, Colorado, on the edge of the Rockies south of Denver. A modest settlement when the General first put in his line, Pueblo had emerged as a thriving trading center with a proper layout of right-angled streets, a few good-sized buildings, some retail shops, plenty of noisy saloons and some serious retail bustle after the Santa Fe arrived to make it two. The two travelers were ticketed to go south for El Moro, that drab coal mining town of the General's at the current southern terminus of his Rio Grande tracks eighty miles down, just shy of the New Mexico border, though they were not traveling together.

One was from William Barstow Strong's Santa Fe Railroad. The other was from General William J. Palmer's "Baby Road," the Rio Grande. Although both were fairly prominent in the world of the western railroads, that world was so thinly populated that the men had never met. But each knew enough *about* the other to be watchful: each was the very last person on earth the other wanted to find aboard that train that night.

The Rio Grande train was most likely pulled by the Las Animas, a mighty locomotive the General had purchased from the Baldwin Locomotive Works. It had been built back East at the company's vast plant in Philadelphia, which took up an astounding eight blocks of the city's downtown. Since it weighed fifty tons, Palmer had the locomotive shipped west in parts to be assembled on the Rio Grande tracks. The fully reconstructed Las Animas had an angled cowcatcher in front to shove any obstacles out of the way as the train roared along, with two spanking yellow domes capping the engine's boiler. Charging across the open prairie, the train made for a roaring, clanking slab of iron, forest green with yellow trim, trailed by a thick and smelly plume of coal smoke. In this largely barren outback, its arrival was excitement itself, the sight, sound, and smell of progress.

Running on steam, the Las Animas put out six hundred horsepower and kept up a steady clip of twenty-five miles an hour. If the train relied on actual, paired horses, they'd have stretched a full half mile ahead. Such a speed also made the train the very devil to stop. The job was left to brakemen who had to climb atop the teetering passenger cars and hand-crank metal brakes down onto the track, the second riskiest task in America after the physical coupling of the cars, which could clip off a man's arm or throw him under the wheels.

All railroads depended on a curious fact of nature: that a cloud of steam is a spectacularly potent force, so determined to break free of any containment that it will drive a piston to escape. This fact lay behind the industrialization that was, in 1878, busy transforming nearly every aspect of human life through its revolution of industry and transporta-

A Denver & Rio Grande locomotive from the period.

tion. The first steam-powered engines pushed steamships across water until an Englishman thought of inserting them onto wheeled vehicles that could roll on land atop straight iron rails that wouldn't freeze, flood, or dry up. An American inventor brought the concept to the United States, developing a darling locomotive called Tom Thumb to do the hauling. To publicize this marvel of engineering, he raced it against a horse—only to lose when a gasket blew. Still, the concept caught on to form the basis of America's first railroad, the Baltimore and Ohio, in 1827. Countless more train lines crisscrossed the East before the first transcontinental finally ventured west four decades later, sounding a steam whistle as it went.

The Rio Grande train lumbering south now had just three passenger cars, each one a "long, narrow wooden box, like a flat-roofed Noah's ark," said a young Robert Louis Stevenson, who rode such a train for a magazine story a year later. There were likely fairly few passengers aboard, so the two mutual strangers from Pueblo could sit well apart, each one settling himself onto one of the tight, uncomfortable wooden benches. Most of the trains of the more established East were far cushier, offering mobile parlors for the well-to-do, but this was the rugged West. While the standard gauge was wide enough to offer four seats across, the Rio Grande's narrow gauge left room only for benches for two on one side of the aisle and just one on the other, an arrangement that was reversed in the middle of the car, for balance. (The narrow cars could still tip alarmingly in the gusts of wind that whipped across the plain. Just the year before, one had toppled over entirely in a gale.) Overhead, hanging oil lamps swung about as the train rocked along on the uneven tracks, jostling the passengers sideways as the train sped forward. A meager coal stove in the middle of each car kicked out a little heat. At either end a "convenience" provided, according to Stevenson, "a somewhat dangerous toilet" on a lurching train.

The tracks ran south along the edge of the craggy Rockies, and they made for quite a divide. To the passengers' left lay the known world—a broad plain, its whitened mesas looming up off the icy desert like tombstones, stretching back to settled Kansas. To their right,

the unknown world—a mostly uncharted, largely uninhabited territory lying behind those menacingly high peaks that glistened in the
twilight.

It is doubtful that either man much took in the view. They were
the chief engineers of their respective lines, and unbeknownst to each
other had both boarded the train for the same urgent purpose: to seize
the Raton Pass, the sole passage through the mountains to the great
Southwest. Each of their bosses, Strong and the General, had independently decided it was key to his future. The General had long thought
so, but not yet acted on it. Strong had long been aware of the General's
interest but come to see its brilliance only upon taking the Santa Fe
job. The task now for each of these men was to claim it—to get a work
crew there first and start digging. On the western frontier, possession
was not nine-tenths of the law but just about all of it.

The chief engineer of any railroad is a key figure, and Strong and
the General had each dispatched his man personally. The two train
men had been tracking each other's movements practically from the
moment they parted at Glen Eyrie, and the situation had turned into
a matter of spy versus spy. All the most sensitive cables were sent in
code—but then each man intercepted the other's and bribed the telegrapher to decode them. The messages conveyed a tightening of focus,
and an increasing need for haste, but they did not reveal the nature or
timing of any clear plan. The whole business induced a rising panic in
the two rivals, but neither was free to act.

The General was in the tighter bind. By early 1878, he'd become
dangerously overextended. His branch lines reaching into the coal
fields of the foothills of the Rockies had not produced the revenues
he'd hoped for, leaving him reluctant to push past El Moro at the bottom of the state to Raton—let alone into virgin New Mexico on the far
side. He'd decided it was safer to wait.

But Strong had his own problems. New to the job, he had difficulty
persuading the somewhat stodgy financiers of the Boston Crowd to
leap into the Southwest, which they felt would just swallow up their
railroad and never cover the costs of construction. Unlike the Gen-

eral, however, Strong lived by that one maxim, Grow or Die, and he saw the Raton Pass as the company's route to prosperity. To him, it was the Santa Fe's destiny to reach the holy city of its name and press on to the sea from there, but to the Boston Crowd, it made much more sense to go straight on through the Rockies from Pueblo, picking up revenue from local mines and then blasting west from there when the time was right.

The General couldn't imagine that Strong would spring for Raton any time soon. The Santa Fe, after all, had only fairly recently reached Pueblo, and for Strong to hit Raton, he'd now have to swerve drastically south and build a lot of track along territory that the General had already claimed for himself. Why would Strong ever want to double the General's tracks, only to fling himself over some distant mountains into the unknown? And so early in his tenure?

But the fact was, from the moment he had joined the Santa Fe, Strong was determined to hit the pass—and to hide his intentions from the competition. The General might have experience as a military spy, but Strong knew a few things about keeping secrets. Shortly after leaving Glen Eyrie that November, Strong had hired a surveyor to plot a train route over Raton, directing him to disguise himself as a Mexican shepherd so he could map the exact contours of the local topography without drawing suspicion.

Such subterfuge succeeded only into February, though, when the General intercepted enough cables from him to Boston to realize that the upstart might be planning the unthinkable. Through his own surveillance channels, Strong likewise sensed that Palmer had become all too aware of his interest and was now responding to it with plans of his own.

In each man, the anxiety went deep. It was as if the needles on their internal compasses no longer oriented to their own intentions, but to the other man's. What the devil was he up to? And this forced on each man a begrudging reappraisal. Previously, each had defined himself by his difference from the other. Now, suddenly, that difference wasn't so clear, for each was starting to act *exactly* like his rival.

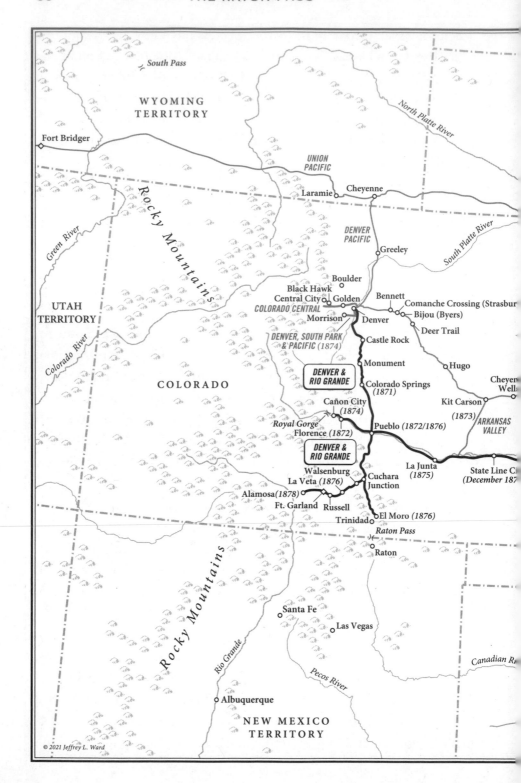

South Pass

WYOMING
TERRITORY

Fort Bridger

UNION
PACIFIC

Green River

Rocky Mountains

North Platte River

Laramie   Cheyenne

DENVER
PACIFIC

South Platte River

Greeley

UTAH
TERRITORY

Colorado River

COLORADO

Boulder
Black Hawk
Central City   Golden
COLORADO CENTRAL
Morrison   Denver

Bennett
Comanche Crossing (Strasbur
Bijou (Byers)

DENVER, SOUTH PARK
& PACIFIC (1874)

Castle Rock

Deer Trail

Monument

Hugo

DENVER &
RIO GRANDE

Colorado Springs
(1871)

Cheyen
Well

Cañon City
(1874)

Kit Carson
(1873)   ARKANSAS
VALLEY

Royal Gorge
Florence (1872)

Pueblo (1872/1876)

DENVER &
RIO GRANDE

Walsenburg
La Veta (1876)

Alamosa (1878)
Ft. Garland   Russell

Cuchara
Junction

La Junta
(1875)

State Line C
(December 187

El Moro (1876)
Trinidad

Raton Pass

Raton

Santa Fe

Las Vegas

Rio Grande

Canadian R

Pecos River

Albuquerque

NEW MEXICO
TERRITORY

© 2021 Jeffrey L. Ward

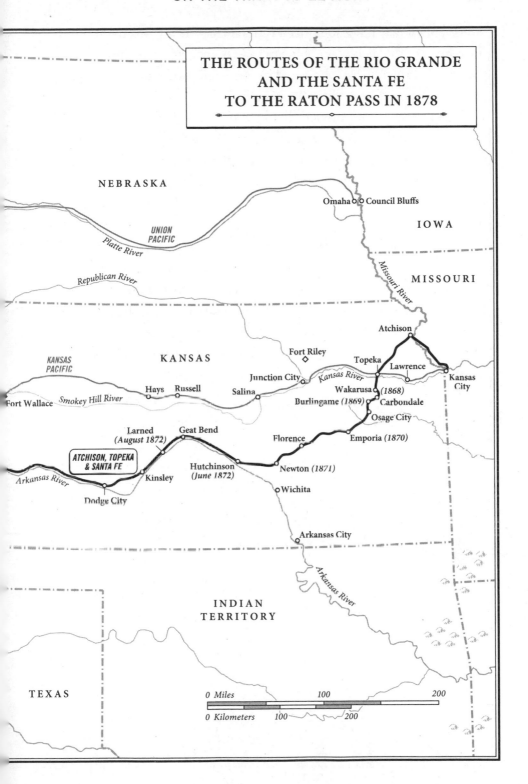

THE ROUTES OF THE RIO GRANDE AND THE SANTA FE TO THE RATON PASS IN 1878

NEBRASKA

Omaha o o Council Bluffs

IOWA

UNION
PACIFIC

Platte River

Republican River

Missouri River

MISSOURI

Atchison

KANSAS
PACIFIC

KANSAS

Fort Riley

Topeka

Lawrence

Kansas
City

Junction City

Kansas River

Wakarusa o (1868)

Burlingame (1869)

Carbondale

Hays   Russell

Salina

Osage City

Fort Wallace   Smokey Hill River

Fort Wallace

Larned
(August 1872)

Geat Bend

Florence

Emporia (1870)

ATCHISON, TOPEKA
& SANTA FE

Hutchinson
(June 1872)

Newton (1871)

Arkansas River

Kinsley

o Wichita

Dodge City

Arkansas City

INDIAN
TERRITORY

Arkansas River

TEXAS

0 Miles   100   200

0 Kilometers   100   200

The General may have convinced himself that Strong was the soulless figurehead of the money-grubbing Santa Fe, Inc., while he himself was the rugged individualist who acted on principle, and Strong may have seen the General as a blithe aristocrat, while he was a businessman with one overarching conviction, Grow or Die. Now as their two engineers rattled through the night to El Moro, the two men's objectives were merging into one. And as they did, the fundamental distinction between these two very different bosses was gradually draining away, leaving each to wonder, once their ambitions collided at the Pass, who he was now.

But, of course, there was more at stake than these men's egos, for they were struggling to determine the direction their lines would take. Initially, of course, the two lines ran perpendicularly to each other, the Rio Grande due south, the Santa Fe due west. But now, as the two lines became entwined, they settled on a new, common direction, running neither south nor west, but into the Southwest on the same route to the sea.

CHAPTER 2

The Wild West

IT IS HARD TO IMAGINE NOW, BUT WELL INTO THE 1870S, most of the American West, for all its cloud-topped peaks and gorgeous coastline, was a barren and uninhabitable tundra as far as most Americans knew. If the United States opened like a book, just about every sign of domestic habitation would be on the right-hand page, and, as if by divine ordinance, the crease down the middle followed almost exactly the 100th meridian that ran just west of the Mississippi. The left-hand page was widely dismissed as the Great American Desert.

Of the 38.5 million Americans counted in the census of 1870, fewer than two million were in the West. (The Plains Indians were left uncounted, but they were never especially numerous.) Of the top ninety cities in America, only two were located west of the Mississippi, and for both their appeal depended on gold. San Francisco, port city for most of the incoming miners seeking gold and a residence for those who found it, edged out Buffalo to claim spot number ten, with a thriving population of 149,473. Sacramento, a mere mining town, came in at number eighty-nine, with a paltry 16,283 even though it was the first western terminus of the Pacific Railroad.

It made sense that the West was still relatively empty. Most Americans arrived in the East, and if they wanted to go farther west, they faced a daunting topography. On maps, the East was colored a lush green, being heavily forested and well watered, while most of the West was a dull brown, barren and parched. The Appalachian Mountains of the East were relatively gentle compared to the towering Rockies of the West, which seemed designed to intimidate. And, for many, the Mississippi served as a moat halting passage beyond. While the many waterways of the East provided easy transportation, there was no equivalent in the bone-dry West. (That function would be filled by trains.) This made America less one country than two.

The United States, of course, was originally made up of the thirteen colonies hugging the Atlantic coast, plus Indian land stretching out to the Mississippi that was secured by treaty from the British after the War of Independence. Another wedge of land, running up from Louisiana through the Midwest to Montana, was added in 1803 as part of Jefferson's famous Louisiana Purchase. The final piece was acquired in a series of bold strokes by President James K. Polk, a one-termer who has never received his due credit. Polk had been determined to fulfill America's supposed Manifest Destiny to reach from sea to shining sea, and he did, acquiring much of the Pacific Northwest in his own treaty with the British. He also formalized the annexation of Texas, formerly a breakaway state from Mexico, and took the rest of the spreading flatlands of the Southwest out to California as the spoils of the American army's sweeping victory over Mexico in 1848. Altogether, Polk added all or part of what would become thirteen western states, a full third of contiguous America.

The US government had been duly sending out explorers to investigate these mysterious western lands, starting with the most famous of them all, Lewis and Clark, in 1803. But as late as 1869, when the first transcontinental railroad was completed, large portions of the West remained the "Great Unknown," as the grizzled, one-armed geographer John Wesley Powell put it. As he paddled down the Colorado River into the otherworldly chasm he named the Grand Canyon, he summed up the mixture of terror and admiration that all these early western explorers experienced:

We have an unknown distance yet to run; an unknown river yet to explore. What falls there are, we know not; what rocks beset the channel, we know not; what walls rise over the river, we know not.

Ah, well! we may conjecture many things. The men talk as cheerfully
as ever; jests are bandied about freely this morning; but to me the
cheer is somber and the jests are ghastly.

It was left to the railroads to dispel the terror of these alien land-
scapes.

▸◂▸◦◂▸◂

The first train to venture into the uncharted West was the Union Pacific, a new railroad company that had been created by Congress solely
for the purpose. Extending out from the Midwest, the UP had linked
up with another federal creation, the Central Pacific, coming in from
California. Together, they created that first transcontinental, the Pacific Railway.

May 10, 1869, was a glorious day for the young country, as it marked
the occasion when the two lines met at Promontory Point in the Mor-
mon country north of Salt Lake City. It was a fantastic engineering
achievement and a triumph of heroic perseverance, especially by the
track layers, mostly Irish going west, mostly Chinese coming east, who
engaged in a spirited competition to outdo the other. (That honor fell
to the Chinese.)

But when judged purely on commercial terms, the line was a dismal
failure. It was the moonshot of the railroad era, an accomplishment
that was only symbolically significant. It was all too telling that when
Leland Stanford, one of the four powerful money men behind the Central Pacific, tried to slam home the Golden Spike with a monstrous
hammer, he missed it entirely. This was especially embarrassing since
the blow was to complete an electric circuit that would automatically
send the thrilling news out over the telegraph wires that accompanied the tracks. A telegraph operator had to key in the word that flew
around the world—"DONE!"—setting off a chorus of bell-ringing all
across the country, led by Philadelphia's cracked Liberty Bell. In San
Francisco two hundred and twenty cannons boomed forth; Washington, D.C., fired a hundred more. Chicago greeted the news with the

Celebrating the completion of the Pacific Railway
at Promontory Point, Utah.

biggest parade of the century. And so it went, total jubilation all around
the country. It was as if these intrepid Americans had discovered a new
continent, come up with a spectacular invention, or won a world war.
And, indeed, they had done all of these. What they had not done was
find a sound business rationale for the endeavor.

A railroad to the Pacific had first been proposed by a New York merchant, Asa Whitney, back in 1844, and was long championed by the
visionary engineer Theodore Judah. But it was the former railroad lawyer Abraham Lincoln who pushed the initiative through as president
in the war year of 1862. Lincoln had also set the width of track for the
railroad, establishing the peculiar distance of four feet eight and a half
inches as the national standard for a train system.* The odd measure
actually stemmed from an obscure English line that may have taken it
from the distance between Roman chariot wheels. It seemed to offer
a usable compromise—narrow enough to be affordable, wide enough

---

* Previously, American trains ran on a maddening hodgepodge of track widths. When
Lincoln went by train from Springfield, Illinois, to New York City to receive the Republican presidential nomination for president in 1860, his trip took four days because of all
the delays in transferring to trains on five different widths of track. If he had ridden a
single width of track, the trip would likely have taken one.

to be capacious. Lincoln saw a transcontinental train as a means to join East and West without the horrifying bloodshed of the Civil War that was then raging. It would have the added benefit of solidifying the north's hold on vast California, which otherwise might have slipped away like Texas to the Confederacy. The project relied on the government to provide the capital, much of it in the form of copious trackside land grants for the trains to develop towns for profit along the way. Such federal largesse was indispensable. For the distance across the West was so great, the tracklaying so expensive, the physical obstacles so many, the threats from Indians so terrifying, and the scale of the enterprise so audacious, no railroad man would have gone along, especially in wartime, if it had been purely a financial proposition. But it also made the Pacific Railroad far more a political act than a commercial one, and that was its undoing—riddling it with the corruption of insider politics while insulating it from the efficiencies of markets.

Financially, it proved to be such a disappointment that, once the hullabaloo died down, many wondered why it had been built in the first place, and, now that it had, why anyone would build another. Winter snows made the line impassable for long stretches of the year; precious few towns created in the vast outback of the West ever developed into anything because so few of them connected to anything; and relatively few people were in such a hurry to get West that they'd spend their life savings on a ticket. Some of these issues were anticipated by its backers, which is why they pitched the railroad less for its utility for passenger transportation and more for its potential use in shipping eastern goods speedily to Asia. But even that failed to materialize.

The investors who'd put their own money into the Union Pacific soon regretted it. Countless bank loans were in default, and, worst of all, the whole venture was revealed to be a scam. The UP backers had created a theoretically independent construction firm, grandly termed the Crédit Mobilier—after a similar scheme in France, in the hope that its exotic Europeanness might be a mark in its favor—to do the actual laying of track, making the Union Pacific itself just a corporate shell designed to reap the profits that stemmed from the healthy differential between the prices charged by Crédit Mobilier and the actual costs incurred.

To create the Crédit Mobilier, and to secure the funding of this boondoggle, the UP backers had paid off seemingly every political power player in Washington from vice president Schuyler Colfax and future president James Garfield on down to push the legislation through. When the facts were revealed in 1873, the nation recoiled in shock. KING OF FRAUDS, bannered New York's *The Sun*, COLOSSAL BRIBERY. One of the men accused of making the bribes, Congressman Oakes Ames (of the Ames Shovel Company that stood to profit from the construction) was mocked mercilessly in the newspapers as "Hoax Ames." Mortified, he dropped dead from a heart attack.

Through the scandal, the American citizenry was introduced to a much more insidious form of corruption, a creation of "friendships" between powerful train men and powerful government officials, based on a sense of shared enterprise that redounded to their mutual financial benefit, regardless of the public interest. That, in turn, revealed something important about railroads as they scaled up to go national. They were becoming nations themselves, with their own economies, customs, laws, and governments. All too often, they engaged with America as sovereign equals, with potentially calamitous results.

That first transcontinental linked up East and West in only the most theoretical sense. To start with, despite its name, it actually traversed only the West. By 1869, there were plenty of lines in the East—several hundred of them having been built up since the Baltimore and Ohio— blackening a railway map with more track per square mile than anywhere in the world. It was telling that the Union Pacific did not head West from Chicago or St. Louis, which were then the nation's western hubs, but from obscure Council Bluffs, Iowa, which would require a separate link to Chicago. Why Council Bluffs? Largely because the

UP's chief engineer, Gen. Grenville Dodge, had grown up there, had substantial land holdings in town, and was friends with President Lincoln, who made the final determination. By the same token, rather than start in San Francisco, the West's largest city, the Central Pacific originated in the much-smaller Sacramento.

There was another reason why the Mississippi had marked the limit of train territory for so long, aside from its role in dividing the lush East from the barren West. Not only was it wide enough that it was not easily bridged, but it was filled with steamships that carried goods north-south and were not keen to see any goods cross the river to go east-west by train. This attitude was expressed unequivocally in 1856 when the fledgling Chicago and Rock Island Railway had the temerity to build a bridge across the Mississippi from Rock Island, Illinois, to Davenport, Iowa. A steamboat coming up from New Orleans deliberately slammed into its foundation, dropping the whole thing into the river in a fiery blaze. Sympathetic river men raised a banner: MISSISSIPPI BRIDGE DESTROYED. LET ALL REJOICE.

When the courts tried to determine liability, the question turned on who owned the river. Was it the exclusive property of river boats— or could it be traversed by trains, too? The future of the nation hung on the decision. When the Supreme Court sided with the trains, the whole country rotated. No longer would America run north-south with the Mississippi and the Atlantic coast. Now it would go east-west with the trains. The lawyer for the Chicago and Rock Island who won this victory? A former Illinois congressman named Abraham Lincoln.

If judged purely in commercial terms, the first transcontinental was an abject failure. But there were other ways to look at it, starting with the geopolitics of the matter and going from there. It was no small thing to rotate America ninety degrees to run east-west. And in the process, the first transcontinental did something even more astounding. It altered the popular understanding of space. Previously in America, distance was defined by how far you could go on foot, then it was

A newspaper cartoon expressing American
joy over east–west unification.

how far a horse could take you, either on its back or pulling a stage.
Those journeys marked the natural limits of your world.

Trains changed all that. The early eastern lines just compressed distance, bringing Chicago, for instance, much closer to New York, once
trains joined them. But when the Pacific Railway came in to speed
people to the far coast, it annihilated any mental idea of distance at all,
in part because the country was so unimaginably wide, and replaced it
with the more comprehensible measure of time. After the first transcontinental was completed, San Francisco was no longer a head-bending
three thousand miles from New York, but a tidy seven days away. Before
long, it would be three and a half.

And then time fell away as the measure, too, for a journey was tabulated instead by what was becoming the universal unit in that hypercapitalist era, its price. No longer was the question how far or even how
long. It was "how much?" Distant places seemed nearer as the price of
a ticket dropped and farther if the price rose. This shift happened in
the West ever more frequently as more trains came in to drive the fare

down—but that price could rise again if a new train drove the old one out of business. The whole matter was even more dizzying for shippers, as the rate charged by a single train might vary with the good to be transported. A ton of shovels might be cheaper to send than a ton of wheat, making the destination nearer or farther, depending. In time, trains charged more per mile for short hauls than for long ones, collapsing or expanding distance accordingly. It made the West strangely intangible. It was no longer under your feet. It was in your head.

For passengers, that was just the beginning. A train was not just faster than anything that had come before but made for a far more pleasant ride. And this was no mean feat, for it required a physical rejiggering of America. The great tumult of the nation's mountains, canyons, rivers, ravines, forests, and deserts—all of it had to be leveled so that trains could go straight along the flat: Anything above a four-percent grade—meaning four feet of rise for every hundred feet of run—was a near impossibility. What God hath wrought civil engineers would now revise. And they did, ultimately running out a vast web of iron rails, more than two hundred thousand miles of them altogether by the end of the century, enough to circle the globe nearly ten times. Construction crews tunneled through mountains, gashed ridges, switchbacked over rises, bridged rivers, and overpassed ravines. All of it so train passengers everywhere could sip tea as they rocketed along.

It was amazing. Even as the trains pushed through the vast, still-forbidding wildness of the West, they provided much of the coziness of home. At the high end, the wealthy could turn their private cars into mobile drawing rooms, complete, as one publisher's wife gaily noted, with "bouquets, shawls, rugs, sofa-cushions," not to mention the "various personalities" of her companions in the "general salon" of her Wagner Palace Car. Before long, General Palmer himself would travel aboard a three-car train all his own, the Nomad, its interior done in deep-red upholstery with mahogany furnishings. The first car held a fully staffed kitchen and dining room, the second offered berths for four guests, and the third was reserved for himself, with a glassed-in observation deck trailing behind.

Those who'd previously traveled by stage or on horseback had to be bowled over by how *nice* it was to go by train. There was no wind in their faces, and none of the stagecoach's lurch or the horse's discombobulation. Inside a train car, a passenger had no sense of the weather beyond a hint of winter chill or summer heat. No one ever got wet, or muddy from rain slop. Outdoor sounds were muffled, and all smells came from oil lamps, a fragrant seatmate, or dinner.

As the early rail passengers sped along, none could fail to notice a highly curious phenomenon that, in its way, superseded all the others—that strange blur of things outside the window. It was like the world had somehow dissolved. Near objects whipped by so fast that they lost their physical integrity and appeared only as streaks of color. So observed Ralph Waldo Emerson back in 1834 when he first rode a train. "Matter is phenomenal"—meaning the stuff of mere impression—"whilst men and trees and barns whiz by you as fast as the leaves of a dictionary," he declared. "Trees, field, fields, hills, hitherto esteemed symbols of stability, do absolutely dance by you." This transition from matter to phenomenon—that was the blur. The train seemed to devour the world it raced through. But, curiously, only passengers gazing out the

The interior of General Palmer's private railway car, the Nomad.

side of the train would see it that way. The train's engineers looking ahead up front had no such impression. And, no less remarkable, the blur did not extend all the way out to the horizon. It consumed only the near objects. Far ones remained fixed to create a distant panorama, a kind of mural in the distance. This was particularly pronounced on the trains of the West, as they sped through wide-open territory that offered a grand vista. This perspective somehow enhanced the grandeur of the landscape, emphasizing its bigness. But it also made it seem like solid reality was beyond their reach.

As innovation proceeded, and trains ran faster, the blur extended out still farther until it threatened to consume everything. But by that time, travelers were so used to this optical illusion, they had long since stopped marveling at it, or even noticing it, particularly. Nine years after he had first been taken aback by the blur, Emerson himself simply noted that it was merely "dreamlike" to see everything whiz past without making any "distinct impression." And he left it at that. He was by then far more taken by the fact that his passenger car glided so smoothly along he could pass his entire journey absorbed in a French novel. He was not alone. Aboard a rollicking stagecoach, to say nothing of horseback, no one could possibly have read a book in any language. But it turned out that train rides were for reading. Indeed, railroads did much to expand the publishing business, both because train travelers provided a new market for publications of all sorts—books, magazines, newspapers—and because the trains could disseminate them more widely. The trains spread the news, and enlightenment with them.

A blur outside, stillness inside—those were the essential characteristics of high-speed train travel. But just as the blur became forgettable, the stillness could be annoying, leaving irritable passengers with the idea they were just packages to be shipped. "I get so bored on the train that I am about to howl with tedium after five minutes of it," the French novelist Gustave Flaubert complained. "One might think that it's a dog someone has forgotten in the compartment; not at all, it is M. Flaubert, groaning." But that was hardly the dominant view.

For many rail passengers, the overwhelming impression was not of riding at all, but of flying. Aboard some projectile, possibly, or perhaps even becoming one. But the exhilaration was in there, too. Freud saw the erotic in the way trains rushed people ahead, unstoppably, in what might have been, in another context, breathless abandon. Indeed, trains early on acquired the quality of romantic adventure, as honeymooners often took off on a train. Affairs could also begin on one. Conversation with a mysterious stranger might reach unexpected depths over the course of a long journey. The romantic possibilities reflected the fun part of being in a land of immigrants, everyone thrown in together to begin life anew.

Anonymity plus intimacy could be a potent combination, turning a train ride into a novel that a passenger could actually enter by taking a seat. And a novel of almost any genre—romance, adventure, domesticity, detection, even horror. Trains offered a rare chance to view the human drama close up, and potentially to participate in it. Everyone was playing a part—the newlyweds, the traveling salesman, the crippled Civil War soldier, the freed slave. It was fun to learn where a seatmate was headed or had come from, and a thrill to get the full attention of an attractive stranger.

But of course such encounters might well *not* be pleasant, too. Initially, passengers had trouble coping with the idea of being private in public, and vice versa, especially in mixed company. It took a while to determine the rules of these social relations. For their part, women tried to always be of "respectable" appearance to maintain some social distance in close quarters with men, and this extended from their attire to their deportment. This was the Victorian era, after all. Any number of magazine articles warned women to be wary of the small courtesies of men, lest they lead to some drastic moral compromise. It was better, wrote the *New York Times*, for a woman not to accept a peanut from a strange man. But this did not keep strange men from offering.*

---

* As it was, men tried things they'd never dare to do closer to home. One popular cartoon showed rows of train passengers behaving decorously—until the car plunges into the darkness of a tunnel, whereupon one man plucks out a bottle of whiskey to suck on

Operating as a single line across the West, with no other lines to feed it, the Pacific Railway was likely to face a dearth in passengers. But this shortfall was compounded by its failure to develop new towns along its route to generate traffic on its own. It was this combination that doomed the first line commercially. Indeed, the town-site selections were often shockingly random. One eastern journalist explained the process: "Some official put an inky finger on the map. 'There,' he said, 'is a good place for a city. Call it Smith's Coulee after our master-mechanic." Too many of these new towns flared only briefly when they were the line's temporary terminus, the end of the line, and then returned to obscurity, if not nonexistence, when the tracks pushed on farther West. These were the notorious "Hell on Wheels" towns that gave the West the uproarious cast that General Palmer sought to avoid, since it did nothing to inspire easterners to move there. A distraught railroad agent decried them in *Harper's Magazine* as "wicked, wonderful and short-lived," temporary quarters for "canvas saloons, sheet-iron hotels, and sod dwellings surrounded by tin cans and scattered playing cards." He mocked them as "air towns" that would soon be nothing but "punctured bubbles."

To be fair, the Pacific Railroad did try to boost some of these new towns but met with only the limited success that comes from inexperience and poor strategy. When the Union Pacific wised up and shifted its eastern terminus from Council Bluffs to the more promising Omaha, Nebraska, ten miles away, it entrusted the promotion of this portion of the outback to the eponymous "Mr. Train"—George Francis Train, an

---

and another tries to steal a kiss from a startled female seatmate. Concerned about the lack of behavioral standards on these "common carriers," the government stepped in to force the railroads to take responsibility for their passengers' well-being. Railroads responded by creating a rulebook on passenger behavior, much of it designed to discourage male drunkenness and lewd behavior. They also set strict rules for employees' conduct, chiefly in regard to their uniforms, from which, said the Northern Pacific "no deviation . . . will be permitted." But of course there was no rule book to govern the manners of the men who ran the trains.

early backer of the Union Pacific who'd had a hand in the creation of the Crédit Mobilier financial device that proved such a scandal.*

Train was enlisted to deploy his considerable promotional skills in boosting the Union Pacific's new railroad towns, starting with Omaha and nearby Columbus, which Train thought he could plump up not just to be the capital of the state, but of the entire nation, supplanting Washington, D.C. In Columbus, though, the closest he came to making it the seat of the federal government was to put up a hotel with the best rooms reserved for the American president. None ever came, but thus the notion of a presidential suite was born.

While that promotional effort may have failed, Train ensured his own success by buying up five hundred acres of Omaha on the cheap— and then selling them for a higher price after it was announced the railroad was coming to town. Once a terminus, the town grew pretty much on its own, adding a new district dubbed Traintown, a play on both the railroad and its promoter. Train's most lasting contribution to Omaha was the 120-room Cozzens House Hotel, which he built out of spite: Apparently, he'd felt insulted by the poor service in Omaha's sole and rather modest hotel, the Herndon. In retaliation, he found eastern backers to erect a far grander one directly across the street to drive the Herndon out of business.

In fairness, Train did bring a focused energy to the project of making Omaha a Paris of the West. One journalist marveled: "He drinks no spirits, uses no tobacco, talks on the stump like an embodied Niagara,

---

* Even in an era of outsize personalities, Train was something of a wonder. For he was also a peripatetic self-promoter later known as the inspiration for the globe-trotting Phileas Fogg in Jules Verne's *Around the World in Eighty Days*. He claimed to have invented the eraser tip for a pencil, the folding carriage steps on horse-drawn broughams, and the boutonniere for aspiring gentlemen like himself. Near the end of his life, Train's lively eccentricity tipped toward frank insanity after he was imprisoned in the Tombs of New York City for publishing a racy article, judged obscene, about the Rev. Henry Ward Beecher's alleged extramarital affair. He emerged a broken man. He took to shaking his own hand when greeting others. Eventually, he marketed himself as a public spectacle, charging admission to rallies in his campaign to become, per his term, "Dictator of the United States."

composes songs to order by the hour as fast as he can sing them, remembers every droll story from Joe Miller to Artemus Ward, is a born actor, intensely in earnest, and has the most absolute and outspoken faith in himself and his future." At a time when standard promotional efforts consisted of flags, a brass band, and a platoon of beribboned dignitaries, Train upped the ante. To boost the train line, he duly gathered two hundred potentates, including Nebraska's two senators-elect; future president Rutherford B. Hayes; George Pullman of the passenger car company; and the full board of the Union Pacific. And he enlisted the Great Western Light Guard Band of Chicago to entertain them before sending them all down the tracks aboard a pair of Union Pacific trains pulled by two flag-decorated locomotives. After crossing the broad and rather spooky Platte River Valley, the entourage camped for the night in "a brilliantly illuminated encampment," according to one attendee, a man named Seymour, where everyone dined under a vast tent and, after dessert, enjoyed a war dance performed by friendly Pawnee warriors. "Of all the wild and hideous yells, grotesque shapes and contortions that have ever been witnessed by a civilized assemblage in the night-time upon the plans," Seymour enthused, "this was most certainly the climax."

Actually, the true climax came the next morning when the Pawnee crept back just before dawn to stage an impromptu Indian attack just for the fun of terrifying the excursionists. Still, Seymour couldn't help being struck by the final tableau as the Pawnee and the dignitaries bade each other goodbye from opposite sides of the Union Pacific track. The "extremes of civilized and savage life," as Seymour put it, face-to-face, although, as the railroad continued mercilessly on, it was less and less clear which was which.*

---

* The Pawnee stunt brought the Union Pacific reams of publicity, but few sales. One writer later declared that of all the places in America none was "so well lied about" as Omaha. The Pawnee warriors who entertained Train's dignitaries ended up performing in the Wild West Show of "Buffalo Bill" Cody.

CHAPTER 3

Beware the
Prairie Lilies

WHILE MR. TRAIN WAS BUSY UP IN OMAHA, PALMER VEN-
tured down to the bottom of Colorado to get his first look at the Raton
Pass. He was with the Kansas Pacific, in charge of plotting its future
route from Denver to the sea. The idea was to take a latitude through
the Southwest below the Pacific Railroad but above the line projected
by the Southern Pacific along the Mexican border.

This was the part of America that would eventually be formed
into the four squared-off states of Utah, Colorado, Arizona, and New
Mexico, but for now was terra incognita. To see what was out there,
Palmer put together a surveying team. Among the applicants was a
young Irish homeopath, Dr. William Abraham Bell, five years Palmer's
junior, whom everyone called Willie. He wanted to be staff photographer, which was quite amusing because, as Bell admitted later, he'd
been entirely "unacquainted" with the art of photography until just
two weeks before. Palmer must have seen through all of that, but he
couldn't help but be charmed by this chipper Irishman with a raffish
beard, a devil-may-care attitude, and countless aesthetic pursuits. It
didn't hurt that Bell's father was now a prominent London doctor with
deep connections to the moneyed end of London high society. That,
Palmer had to think, might be useful to him in if he wanted to start his
own railroad. Palmer signed Bell right up.

Willie Bell.

Bell spent the better part of the next two years riding about five thousand miles of the Southwest, taking photographs of the strange, nearly lunar outback while the rest of the crew hauled about surveying—rods, chain lengths, levels, and the newfangled theodolites—specialized telescopes for sighting contour lines. As he journeyed, Bell was stunned to discover that much of this new territory was "almost without tillage or inhabitants." Every few weeks, they might come across a mining camp, a floundering utopian community, a subsistence farm, a Spanish hacienda, or a military outpost, but that was about it for civilization.*

For this trip, Palmer had made a point of sending his team into the Raton Mountains. He knew that the fabled Santa Fe Trail from Kansas had crossed through there to reach the city of Santa Fe, long the com-

---

* His photographs convey the full shock of it. Printed with round, dinner-plate borders, somewhat blurry, they represented the view through a spyglass, and reveal a lunar vacancy in every direction. A stream, a mesa, a clump of trees, a mountain—those were the sights of the great Southwest. Taken together, they make for a barren, lonely expanse of nothing. One view is captioned "City of Rocks," which sums up the desolation of all of them.

mercial high point past the Rockies. An enterprising Kentucky trader, William Becknell, had followed the traces of Spanish conquistadors to find that rare mountain passage in 1821, and ever since any number of midwesterners had followed the route on oxen-drawn Conestoga wagons, or on horseback, or trudged along wearily on foot. (Becknell's next trip revealed the full hazards of a long trek across the baking desert: He survived by drinking blood from the ears of his mules.) It took a few months for Bell's surveying party to reach the pass. It was about seventy-five hundred feet up among the last of the great hulking Rockies, its grassy slopes fringed by piñon and cedars and shadowed on the east by flat-topped Fishers Peak, a ten-thousand footer that loomed over the valley. The formidable Spanish Peaks rose up farther to the west.

A botanist among his other pursuits, Bell was off hunting wildflowers while the surveyors busied themselves under a wind-battered tent. So he wasn't there when a war party of Arapaho, all of them adorned with feathers and slathered with face paint, loomed up in the distance. Seeing them, the surveyors anxiously sent out their guide to assure the Indians that they were just passing through. The warriors were guileful in return, telling the guide they'd come to fight only the Utes—not any white people. In fact, the Arapaho had come to *recruit* the Utes to fight white people like these very surveyors. The guide was all too well aware that the Arapaho had just plundered and torched some settlers' ranches nearby.

As the Arapaho and the guide faced each other down, doubts mounting on both sides, Bell wandered back to discover the alarming set-to. Happily, one member of the party had thought to send off a messenger to a nearby company of US Cavalry. Several dozen soldiers soon galloped up, guns blazing, to chase the Arapaho off before they could cause any trouble.

Once the danger had passed, Bell returned to his wildflowers. He was intrigued by some particularly entrancing prairie lilies, wispy four-foot plants that overspread several of the pass's boulders and sketched them with his usual delicacy for his travel book. After supper, he spread out a bedroll to sleep among them, since their ghost-white blooms made, he wrote, for "a perfect fairy land."

What he did not realize was that their scent was lightly toxic. Sleeping so close by, Bell drew in the smell too deep, and came down with a wicked headache by morning. So he may have thought he was hallucinating when he spotted Palmer riding up with his private secretary, Captain C. F. Colton, an energetic young man who'd been Palmer's adjutant in the war. Palmer explained had that he'd intended to follow the survey work from behind a desk in St. Louis. But after he got Bell's early reports, he got an itch to see this remarkable landscape for himself.

Palmer hadn't seen Bell since hiring him in St. Louis several months before, and it made for a spectacular reunion out in the wild with so many adventures to report on both sides. Palmer, Colton, and Bell shared a tent for the next few nights, one that Bell wisely moved well away from the prairie lilies. The visit made for such a riot of talk and laughter that Bell declared that, even though Palmer was his boss and Colton just a lowly employee, the two of them were his two best friends in the universe.

To the General, Bell was a cheerfully eccentric Brit who never missed five p.m. tea. And a remarkably erudite one, too, seemingly knowing everything from geology to Greek. Having never attended college himself, Palmer envied such deep learning, but he also shared Bell's taste for adventure, and he relished Bell's many tales of the "red-skins," and not just the close call with the Arapaho the previous afternoon.

Bell had had another encounter, he told Palmer, that he simply could not get out of his mind. He'd been at Fort Wallace, east of Denver, some months before when he heard that a small cavalry company had just been ambushed by hundreds of Indians who'd slaughtered the soldiers to the last man. Ever curious, Bell ventured out of the fort to see what had happened. When he came upon the scene, he discovered the Indians had stripped the corpses of the dead, seemingly to show off the hideous wounds they'd inflicted. A bugler was pierced everywhere with arrows, one soldier had been brutally scalped, another had his arm hacked off.

Chilling as all that was, Bell would never get over the sight of a third, a Sergeant Wylyams. The trampled, bloodstained ground all around him revealed Wylyams's frantic efforts to defend himself as the Indians converged on him. His naked body told what happened next. Wylyams had been shot through the head; his body pin-cushioned with arrows; and his skull smashed in by a tomahawk. When Bell looked closer, he could see other markings on him, grotesque ones that seemed strangely communicative. Wylyams's nose had been slit cross-wise; his throat cut ear to ear; his chest hacked open to expose his heart; his right arm slashed to the bone; his thighs gashed longwise to the knees; and the skin of his forelegs peeled off.

Puzzled by all these cuts, Bell asked around inside the fort afterward to find out if they somehow meant something. He learned that each cut bore the signature of the tribe responsible. The Cheyenne—or "cut arms"—carved the sergeant's bicep; the Arapaho—or "smeller tribe"— slit his nose; the Sioux sliced his neck as well for "the cut throats." So it went, cut by cut, tribe by tribe. The corpse was a letter, written in blood, and signed by each tribe of Indians responsible. It said, *we are not one tribe, but many, and we are united in war against you.**

Palmer must have found that quite disturbing, but he remained undeterred. He would explore this territory for a railroad, whatever the hazards. After bidding Bell a heartfelt goodbye, he left with Colton to poke around the Arizona territory, hoping to find a level passage to California, safely away from the mountains. There he had an intimate encounter with Indians all of his own.

He'd heard that some Apache might be about, so the General had taken the precaution of securing an escort of a small detachment of army soldiers, and they accompanied him when his search took him down into a long, deep, and rather forbidding canyon. Its floor was

---

* The Indians were not alone in seeing the Americans' western expansion as a war against them. Given the the Union Army's role in providing military protection to the transcontinentals, it is not too much to say that the Civil War had continued on after Appomattox, with the Union effort shifted from subduing the rebels of the South to exterminating the Indians of the West.

so littered with boulders that the men had to pick their way through on foot, tugging their weary pack mules behind them. On either side, Palmer later wrote, the canyon walls rose up like the insides of "a great grey coffin." Indeed, one of the mules had lost its footing as it descended and plunged a hundred feet to its death.

Up on the surrounding desert, the General could tell that Apache had been there some time before but, not seeing any, he assumed they were all gone. Suddenly, as he led his men through, gunfire rang out from the lip of the canyon far above, bullets clanging off the nearby rocks. Flights of arrows rained down, too. Then a "dreaded war-whoop" echoed through the canyon. Palmer and Colton dashed for cover behind some bushes while the other men scattered. Then came the terrifying rumble of a rockslide crashing down upon them. The Apache had pried great boulders loose to send them tumbling down the cliff, bringing more rocks down. No one had been hit, but Palmer knew that it was only a matter of time. Peering up the chasm, Palmer could see dozens of Apache warriors exulting atop the canyon's edge.

But the General had been in a few tight spots in wartime, and he quickly thought the matter through. It was eight miles back to the start of the canyon and even farther to the end of it. They'd never make it either way, not with everything coming down on them. The only way out was up—up the canyon walls to take the fight directly to the Apache, no matter how bad the odds. The men would all somehow have to clamber up the steep, slick canyon walls while avoiding the boulders, rocks, and various projectiles blasting down on them. And then, if they succeeded in reaching the top, they'd have to assault all the Apache from below—while being drastically outnumbered and exhausted from the climb. Such a move was more than just brazen. It was suicidal. But it was their only hope.

With brisk shouts and a wave of his arms, Palmer directed some men to climb up the far wall while a few others covered them with rifle fire. Then he sent other men up the wall on the near side, while another group shot off their rifles to protect them.

Palmer himself would go up first, scrambling up the rock face into a hail of bullets and arrows, plus the occasional boulder thundering down. It was a desperate, almost foolhardy gambit, but neither he nor any of his men was struck, and they all kept on climbing, handhold by handhold, foothold by foothold, up the canyon walls. Palmer was the first to reach the top. Keeping low, his pistol out, he scrambled up onto level ground. He could scarcely believe it: No one was there. The dreaded Apache had all vanished, every one of them. He later wrote to Bell:

> *How we got up, God knows. I only remember hearing a volley from below, shots from above, Indian yells on all sides, the grating roar of the rumbling boulders as they fell, and the confused echoing of calls and shouts from the canyon. Exhausted, out of breath, and wet with perspiration, boots nearly torn off and hands cut and bleeding, I sat down on the summit and looked around. Everything was quiet as death; the Indians had disappeared—melting away as suddenly and mysteriously as they had first appeared.*

To the Indians, the threat was not just the white men, but the ideas they brought with them.

The Indians found it simply incredible, this white man's notion that land, the precious earth, could actually be owned by someone. And somehow it was proved by a piece of paper? To the Indians, that was as absurd—and as threatening—as claiming an exclusive right to the sunlight or the air, and it needed to be repelled by force. To the white men, of course, the whole concept of property rights was self-evident, and the basis of capitalism, law, and civilization. They would impose that philosophy on the West, and then exploit it for all it was worth no matter how destructive it might be to the Indian way of life, or risky to themselves.

Willie Bell's write-up of his western travels with the surveying team, *New Tracks in North America*, which proved such a sensation when it

appeared in early 1869 that it caught the attention of Lucien Maxwell
of the Maxwell Land Grant, which lay on the far side of the Colorado
border with New Mexico. Easily the largest private estate in America,
it was twice the size of Rhode Island and contained much of an entire
mountain range called the Sangre de Cristos. It showed the stakes the
two railroad lines were playing for.

Lucien Bonaparte Maxwell was a lively if sometime scatterbrained
personage who blew off a thumb when he lit an antique howitzer to celebrate Independence Day and the gun exploded. (It could have been
worse: The cannoneer lost an arm.) He had started out as a hunter
for the extraordinary John C. Frémont when that future presidential
candidate was still in the explorer phase of his multifaceted career. Frémont's guide for his explorations was the legendary mountain man Kit
Carson.

The history of the Maxwell grant said a lot about the inchoate nature of the West. Maxwell insisted it was all his as the possessor of a
deed that had been bestowed by the Mexican governor on a pair of
his cronies, and then acquired by Maxwell in two parcels. The first
by marrying the daughter of one crony, and the second by buying the
other one out. Nonetheless, the title was so clouded, it would take five
separate rulings from the United States Supreme Court to sort everything out. In the meantime, Maxwell carried on as if the land was all
his. For years, he lived on the estate in an extravagant adobe mansion
that stretched the length of a city block and in a high style, said one
awestruck visitor, "akin to that of the nobles of England at the time of
the Norman conquest."

By 1869, when Bell's book came out, Maxwell was sick of managing
it all. The property was overrun by impudent homesteaders; the mines
were sputtering; and the once-peaceful Indians were acting up. Overwhelmed by it all, Maxwell turned to drink. His once-glorious mansion
became a place for high-stakes poker games, "dusky maidens" and a
"carnival" atmosphere, according to a writer friend. He was looking to
sell, and he thought that Bell might be just the man to write up his vast
holdings in an alluring fashion to entice Londoners to buy a vast piece

of property without seeing it. Such a vast expanse of land like that needed a train—and a train needed such a vast piece of land.

After his survey trip, Bell helped Palmer sell Kansas Pacific bonds to Europeans. He had a knack for marketing, it turned out. He could see that, however appealing the land might be for farming, grazing, and mining, those attributes would not command top dollar unless a train came through. After Maxwell got in touch, Bell saw his chance: He would market the Maxwell Land Grant as the Maxwell Land Grant *and Railway Company*, emphasis added, complete with track rights. Sure enough, Bell sold the property to an investment syndicate headed by a London friend of his for a princely $1.35 million. Bell made sure that Palmer could buy a few shares, too, and then he pitched the General as the syndicate's man on the ground as the Maxwell Land Grant and Railway Company's president.

Once in office, the General invited fellow members of the syndicate to back his nascent plans for a "North and South" railroad (which he now termed the Imperial Pacific in hopes of appealing to these moneyed Englishmen of the syndicate). One buyer was so taken by the train that he sold his $300,000 investment in the MLG & RC to make his purchase, a move that probably did not please his syndicate colleagues. That was the start of the Denver & Rio Grande.

CHAPTER 4

▶─◆─○─◆│◆│◆─○─◆◀

The Grid

IN VYING TO TAKE THE RATON PASS, STRONG AND PALMER were competing not just to lay track through it, but also to develop the property alongside the track they laid. A railroad was not about transportation nearly so much as it was about real estate. It lived off the land it traversed, buying low and selling high once it had put a train through to increase its value.

Early on, the deal was even better, since the western railroads were given land for free by the federal government in exchange for laying rails through it. Overlooking the inconvenient fact that it may not actually have been the government's to give, that made for a massive transfer of theoretically sovereign territory into private hands. Between 1850 and 1871, the United States gave to the railroads almost ten percent of its public domain—land to which no one could make a valid legal claim. It amounted to an astounding 131 million acres altogether. If all that land were gathered into a single state, it would be the third largest in the country after California and Texas. Of the federal grants, the Santa Fe got three million in Kansas, but the federal program ended in 1871 when the government decided, sensibly enough, to let the railroads fund the railroads. So Palmer's Rio Grande got none, and Strong's Santa Fe got no more, much to the two men's frustration.

The initial massive handover of public land to the railroads did much to establish the West as their corporate province, but the manner of it possibly did even more, as it determined the layout of the railroad towns that the railroads built as they proceeded. The federal land was doled out in an odd checkerboard pattern of alternating one-mile squares that extended out ten miles on either side of the tracks, with the black squares of the checkerboard given to the railroads, and the white ones kept by the government. On a map, the tracts of railroad land snaked through the unclaimed territory like a fat river, exactly twenty miles across at every point, three times as wide as the Amazon. What's more, it was a fat river comprised of a patchwork of perfect squares.

If the train tracks imposed straight, flat lines on a turbulent land-scape, the squares that traveled along on either side spread that ge-ometry wider, taking to two dimensions what the tracks had done in one. They brought a surveyor's perfect right angles to a curvaceous, unbounded landscape, and the square, in turn, became the chief char-acteristic of virtually all the towns the railroads built on their land, however it was acquired. The grid.

The railroads did not invent the grid, but merely imported it from the East and then set it down everywhere they went, imposing Car-tesian order on an unruly landscape. The grid dictated that broad avenues marched like a military parade straight through town, with narrow streets regularly crossing them at exact right angles. Foreign as that geometrical concept might have been to the untrammeled West, where people had always set their crude houses down in loose, disor-dered clumps, it was a standard feature of every railroad town past the Mississippi. From Leavenworth through Phoenix, hundreds of towns that were the pure product of the railroads were all physically just the same, block after block, all of them consisting of perfect rectangles under the dome of the western sky, regardless of which railroad made them. Practically as soon as the grid went down, a town went up. A mining engineer from France was in Cheyenne when the Union Pacific arrived, and he could not get over how quickly things sprouted.

Everywhere I hear the sound of the saw and the hammer, everywhere wooden houses are going up; everywhere streets are being laid out,

cut on the square and not an oblique angle as in Europe. There is
no time to hunt for names for these streets. They are street number
1, 2, 3, 4 or A, B, C, D . . . Houses arrive by the hundreds from
Chicago, already made. I was about to say, all furnished in the style,
dimensions, and arrangements you might wish. Houses are made to
order in Chicago, as in Paris clothes are made to order at the Belle
Jardinière. Enter. Do you want a palace, a cottage, a city or country
home; do you want it in Doric, Tuscan, or Corinthian; of one or two
stories, an attic, Mansard gables? Here you are! At your service!

However they acquired the land, all the trains were determined to
turn it into house lots—and the house lots into cash. Luckily, they had
the perfect delivery system to speed building materials to building sites,
and to generate customers in quantity. The Santa Fe and the Rio Grande
went west from Topeka, and south from Denver house lot by house lot.
Everywhere they went, they left their stamp, perfectly right-angled.

Like so much else in America, the grid came from England, specifi-
cally from London, although it seems to have originated with the arrangement of Roman armies at the time of the Caesars. It was brought
by the early American colonists first to Philadelphia, which was arrayed
at right angles, a miracle of squared-off rationality on the banks of the
curving Schuylkill. Soon, the grid spread throughout the industrial-
ized Northeast, landing most noticeably in New York, where numbered
streets and avenues brought tidy geometry to a bumptious city. Eventually, it became ubiquitous almost everywhere outside the agrarian
South, which retained much of its haphazard backcountry character
from its farming days.

The town's plan was all set down in a "plat" that was drawn up
long before anyone lived there, and it rarely paid much heed to the
contours of the actual landscape. In the East, towns obeyed the rudi-
ments of geography. Growing up slowly over time, they followed the
rivers or oceans that had spawned them, curved respectfully around
hills, yielded to mountains. Not in the West. No natural features ever
altered the basic plan. Even if a river sliced through a town, the town

remained resolutely squared up, as if to pay nature no mind. Any house lots that would be submerged were simply deleted. But the railroads regarded *themselves* as the rivers to which the habitations were obliged to conform. And that was nowhere so evident as in the West, with precious few navigable rivers, and fewer still that spawned a town. The railroads did that, instead.

So the surrounding topography was ignored, often shockingly. In 1873, a director of the Northern Pacific rashly hired the landscape architect Frederick Law Olmsted, the celebrated designer of New York's Central Park among countless other outdoor achievements, to create the layout for the Northern Pacific's new railroad town of Tacoma, Washington. Olmsted did the unthinkable: he created a plan for the town with curvilinear streets that rippled out from the rounded hillside. Seeing the plan, the railroad's real estate arm, the Tacoma Land Company, would have none of it. "There wasn't a straight line, a right angle or a corner lot," wrote one appalled critic. "The blocks were shaped like melons, pears and sweet potatoes." The Olmsted scheme was rejected and replaced by a plan drawn up by a railroad engineer that had nothing *but* straight lines, right angles and corner lots.

Much of this simply made sense. Since buildings are right-angled, so should their lots be, and so should the roads that served them. But as settlers spread farther west, they brought this disposition with them, making this vast new country of the West, all of it so different— mountains here, deserts there—everywhere the same. In their uniformity, these instant towns had the quality of mass-produced industrial products not unlike the train cars that were delivering the citizens to fill them up. Initially, they were the air towns derided in *Harper's*, but those were simply the ones that didn't catch on. Plenty more did as time went on, and promoters proved more skillful in divining the desires of potential customers, and developers better at fulfilling them.

The standardization was the attraction, commercially speaking. One didn't actually have to see a town to visualize it. Or to visualize a block, or a building on that block. Such a property could be sold before it existed and resold if need be. If one business failed in a particular space, another could pop right in. Expediency was the point. It instantly turned land into cash for the trains that lived on money.

In the East, the grid had a democratizing effect, as any intersection could become a place of interest, but the railroad towns were hierarchical, as everyone knew where the center was: the railroad station, the landmark that had brought the town to life. Everywhere, these impressive train stations were the biggest buildings in town, and they loomed as the cathedrals of the prairie, the proud symbols of the noisy, thriving commerce, most of them eventually topped by oversize clocks that were the new crosses. Churches themselves were usually shunted to distant corners of the town on the least valuable real estate, if there were to be churches at all. In Cheyenne, services were held in the saloon.

The tracks usually ran smack through the center of town along what was often brazenly called Railroad Avenue. The various railroad buildings—not just the station, but warehouses and offices, too—were placed at the epicenter at the intersection with its central artery, often called Main Street, that ran crosswise. This created the town's epicenter. X marked the spot.

The truth was—a train *did* bring a town to life, and then sustained that life. All newcomers arrived at the station, as did all news, whether it came in via newspapers, letters, or the telegrams that zinged in by wire to the telegraph office invariably located inside. Here too, arrived all the goods from the outside world—hardware, farm tools, crop seed, fashionable dresses, pianos, prefabricated houses—that filled out the inhabitants' existence, turning lives that before had been meager, homespun, and primitive into ones that were ample, provident, and advanced. They were located in the West, but they were furnished by the factories of the East.

While the train tracks created the town's fundamental axis, it also created its social divide. The "other side of the tracks" was not just a metaphor. One side had the banks, the other the saloons. In forming the town's orientation, the tracks determined the path of light and shadow, too. Most western trains ran east-west, so the sun followed the railroad tracks as it cruised across the sky. But whatever the direction, the tracks determined the town's essential compass points, for all roads ran either in parallel along the avenues or perpendicular to it along the streets, gathering the town's blocks into those exact rectangles. In most cases, each block contained precisely six residential houses fronting the street, on lots that ran exactly and invariably fifty feet wide. No streets in a railroad town ever were allowed to meander; the message everywhere was, stay straight. This was the power of the railroads, and the money behind them.

CHAPTER 5

▶─◆─○─◀◆▶─◆─○─◆◀

Where to Go

BUT WHERE TO RUN THESE TRACKS, EXACTLY? NOT JUST anywhere. It might seem to be an advantage for a railroad to be first into a new territory as a monopoly, but that meant being the first to plunge into a void, with no obvious points of orientation. Happily, the Rio Grande's northern terminus of Denver would draw traffic coming down from Cheyenne on a link to the Union Pacific and out from Kansas on the General's old Kansas Pacific. That made for one fixed point. As for another, there was Pueblo, but it was just a dusty trading post along the old Santa Fe Trail. And that was basically it for the whole state.

In a vacant wilderness, it was never clear which way to go. In projecting a line, a train man had to make certain assumptions about the future, but who knew which ones would hold? One might veer a line off to what looked like prime agricultural land to feast off its harvests— only to discover the soil was poor, or, worse, the locusts came every three years. Or give the mountains a wide berth—only to miss out on the gold strike of the century. Plus, there was the physical terrain to consider. Trains wanted to go straight along the flat, but the landscape rarely obliged. Curves, rises, rivers—all of them drove up costs, sometimes prohibitively. How much was too much? How far off course was it worthwhile to go to avoid an incline? What width of river was too costly? How much was a particular town site worth?

The standard text on the topic from the era, *The Economic Theory of the Location of Railways*, published by the railway engineer Arthur

Mellen Wellington in 1887, runs 1,037 pages, and it begins by address-
ing the essential question, placed in all caps for emphasis, "WHETHER
OR NOT TO BUILD THE LINE AT ALL." The rest of the book pon-
ders all the myriad elements that go into that choice—and then loads
on more if the answer is yes. The factors were legion and all of them
measured in hair-fine measures: the level of locomotive power, the degree of track curvature, the price of coal, the weight of that coal, the
availability of railroad ties, the regularity of maintenance, the quality
of available staffing, and a thousand more. And those were just the
costs that could be tabulated. There were plenty more unmeasurable
ones that were more pertinent for the General, like the sheer glory of
being first.

Wellington's tome, with its endless tables and diagrams amid all
the pages of dense, exacting, joyless prose makes for burdensome reading. In the end, one is impressed not so much by all the information
that the inexhaustible Wellington has brought to bear on the topic
as by the desperate need for certainty that lies behind it. In the race
of the known to overtake the unknown, the unknown was still way
ahead. And yet, everything hung on each decision the General made.
To Wellington's way of thinking, to venture south from Denver was to
court disaster. "It is of little avail to run a line even from a great city
to nowhere," he declared. "Without a good traffic-point at each end of
the line, the conditions for great prosperity are not present." (By "great
city," Wellington meant a New York or a Chicago, certainly not a Denver.) Any territory south was indeed nowhere. Forget it.

Nonetheless, starting in 1870, down and down the General went.
This was long before the Santa Fe arrived, when he had all of Colorado
to himself. He'd signed up Bell to handle railway construction, and the
former territorial governor, Alexander Hunt, to make the land purchases. Born in New York, Hunt had come west to make a fortune in
the California gold rush of 1849, and did make one, only to lose it one
night in a poker game. Although he had to retreat to Illinois afterward,
the whole escapade confirmed the westerner in him, and he drove an
ox team from Illinois to Denver to get in on the next great western gold

rush, at Pikes Peak ten years later. This time, he held on to his winnings, and demonstrated enough grit to become a federal marshal. He
soon went up against Jake Slade, a terrifying desperado who adorned
his watch chain with the dried-up ear lobes of his vanquished enemies.
Hunt ran Slade out of the territory, never to return. If Willie Bell was
dreamy, Alexander Hunt was tough as nails.

The three of them, Palmer, Bell, and Hunt, set out to decide what
towns went where. Coming south from Denver, they started with Colorado Springs as the site of a resort to draw vacationers from the East.
Below that, the General basically had two choices: to veer to the southwest toward the lucrative coal fields of Cañon City in the foothills of
the Rockies, or to plunge on straight down to Pueblo. So, which? To
go by Wellington, he'd have to weigh the potential for retail growth
in Pueblo against the possible coal revenues in Cañon City. But the
fact was, that was an impossible calculation. So Palmer simply put the
question to the residents of those two localities: who would pay more
for a train?

In Cañon City, voters offered $50,000—a sum issued in the cur-
rency of Rio Grande bonds.

Now, how about Pueblo? Its citizens doubled Cañon City's bid, offering to buy $100,000 worth of bonds. The General replied that was
a very generous offer, but had to add that Pueblo was a lot farther from
Colorado Springs . . . He got $50,000 more.

In June 1872, Palmer duly delivered Pueblo its train, the first one
stocked with dignitaries for a big banquet at the town's new courthouse.
The contract required the General to place the train station within
a mile of that courthouse, but Palmer blew right past that mark and
put it across the river in a brand-new locale he brazenly called South
Pueblo, although New Pueblo might have been more accurate. If the
merchants of Pueblo wished to take advantage of Palmer's train, they'd
have to move their shops onto fresh South Pueblo lots sold to them by
the General.

It was irksome, but South Pueblo was born in an instant. "Unher-
alded and almost un-thought of, [South Pueblo] is moving forward to

commercial prosperity with the force and momentum of an avalanche," wrote one resident. A few weeks back, he went on, a citizen of Pueblo saw almost nothing across the river. "He now is surprised to behold roofs of dwellings and broad, well arranged streets, while his ears are assailed by the din and clatter of saws and hammers."

His line complete, the General returned his attention to Cañon City once more. This time, he bypassed the residents to appeal to the coal operators. Having seen what a train did for South Pueblo, what would they pay him now? More than $1 million, it turned out. Done. This time, the General turned the construction over to a new concern, the Canon Coal Railway Company. But a year later, its train reached only to Labran, nine miles outside of town. The General told the coal operators in Cañon City that, alas, his train could go no farther because the Canon Coal Railway Company had simply run out of money and he'd need another $100,000—paid in bonds *and* land this time—to finish the job. Indignant, the coal men refused to pony up. But this time, Palmer was not being unreasonable, for this was 1873, the year of the Panic. Trouble had come from the East.

The western trains didn't just build America *out*. They built it *up*, raising America into the industrial colossus that was well on its way to succeeding the British empire as the mightiest in the world. By 1873, the railroads had succeeded farms to become the nation's largest employer, the repository of most of its capital investment, the near-total basis of the stock market, and the creators of the most spectacular private fortunes the world had ever seen. The nation's preeminent railroad man, Cornelius Vanderbilt of the New York Central and other railroad holdings, was well on his way to accumulating a fortune of $100 million, requiring a new word, *tycoon*, to describe someone so unimaginably rich. Many more tycoons would follow.

The speed of the transformation was simply staggering. The Northern money that had previously been bankrolling the Civil War shifted to building trains. By 1873, the total railroad investment had tripled after the close of the war to $3.7 billion, taking the total number of train companies operating in the US to an astounding 364. They pulled the entire economy along with them, raising the number of businesses in America by 50 percent in one year, 1870, alone.

No one demonstrated this shift—and its hazards—more than the financier Jay Cooke. He had been a major player in financing the war effort, dispatching thousands of salesmen into the northern countryside to sell $1 billion in war bonds to villagers who wanted to do their bit. Now that the war was over, Cooke switched to selling Northern Pacific railroad bonds on a similar basis, creating a bank in Philadelphia as his repository. The Northern Pacific had been created by Congress as a second Union Pacific—a private corporation relying on federal funding—but it suffered from the same flaw, much magnified. If there had been little immediate market for the lands of the Union Pacific, there was even less for the lands of the Northern Pacific that ran farther up along the chilly outback of the north. Tracklaying went so slowly, and the returns were so meager, that the company was still a thousand miles short of completion when the Crédit Mobilier scandal broke in 1873, exposing the Northern Pacific's massive vulnerabilities.

Alarmed, the partner who ran Cooke's New York City branch frantically shifted his holdings to his wife's name to preserve his fortune, then shuttered the bank to keep other Northern Pacific investors from retrieving their funds. Cooke then closed the main Philadelphia branch, causing the big bag of air that was the Northern Pacific to suddenly burst. "If I had been struck on the head with a hammer, I could have not been more stunned," said one Northern Pacific executive. The Cooke bank's collapse sparked a run on banks throughout the East, driving forty of them into bankruptcy, and shaking financial institutions everywhere. The president of the Bank of California killed himself when his bank collapsed. Five thousand businesses went under, taking $250 million in debts with them, dragging down lenders and driving up national unemployment to fourteen percent. A "mad terror" so convulsed the stock market, it had to close for ten days. Western Union stock dropped by half, railroads as a class by a third. A quarter of them, eighty-nine in all, went out of business.

In the past, there had been regular economic "panics"—the word for financial disruption—but they had been relatively brief. This one, the Panic of 1873, extended all the way through 1879. For its length, severity and sweep, it would rival the Great Depression as the greatest financial catastrophe in American history. This was the downside of the spectacular railroad boom: while both the railroads and the country grew together, they shrank together, too. Duluth, Wisconsin, had largely been created by the Northern Pacific, and when the railroad went into bankruptcy, Duluth became a ghost town, its population plunging from five thousand to thirteen hundred as refugees left to hunt for work elsewhere. The ruin stoked fury all along the railroad routes, culminating in the biggest job action in American history, the Great Railroad Strike of 1877, when eighty thousand railroad employees went out across the country, and a half million other workers followed in sympathy. In Pittsburgh, Pennsylvania Railroad strikers set fire to the roundhouse, igniting the train station and starting a conflagration that burned down three square miles of the city.*

▸⊷◦⊷◂

Not only did the Panic stop Palmer's line into Cañon City, it put a halt to the Santa Fe move into the state, too. It staggered across the border

---

* It started with the Baltimore and Ohio on July 16 of that year when a fireman, infuriated by a drastic pay cut, stepped off his job at a switching station outside Baltimore, and several other B&O workers followed. The strike moved on to Pittsburgh, Philadelphia, and Reading, then out to the Midwest, where the future labor organizer Eugene Debs joined in at Terre Haute, and ultimately reached as far as Chicago. By the time the strike was over ten days later, nine governors had called out federal troops and eleven had summoned state militia. The conflict closed down two-thirds of the nation's railroad network, put several cities on military lockdown and left 117 people dead.

Pittsburgh on fire in 1877.

from Kansas into Colorado that year but went no farther. Railway construction was locked up everywhere.

To stay afloat, the General pushed up ticket prices on his Rio Grande to over ten cents a mile, more than three times the comparable rate in the East, and he ran freight rates so high that merchants were starting to skip the new railroad altogether to hire Mexican drivers for old-fashioned bull teams to haul their wares. Those extortionate prices didn't win him any friends, and Palmer's method of pitting one town against another was even more irritating. A certain "FBH" wrote in to the *Rocky Mountain News* to declare that the Rio Grande "shares the hate of all Southern Colorado." And in the Colorado Springs *Gazette* an "Interested Spectator" called the Rio Grande "a leach sucking the life blood" out of Colorado Springs.

The *Gazette's* editors rushed to Palmer's defense, pointing out that, having built the town, he was its "surest protector." And the truth was, the General's train had brought prosperity. The four Colorado counties reached by the Rio Grande had seen their property assessments more than double, from $7 million to $18 million after the train arrived. In Pueblo alone, the train had caused 185 new buildings to go up, collectively worth more than $600,000, a very big number on the prairie. Back when Pueblo was served only by stagecoach, only three people a day came into town. Now the train daily delivered a hundred.

Although the national economy would struggle for years more, the first signs that Colorado was finally easing out of the Panic came in 1875. By then, coal men in Cañon City grew tired of seeing their town wither without a train, and they decided to pay Palmer what he asked. Sure enough, once the Rio Grande reached Cañon City it blossomed.

Now Palmer turned his attention to Raton. Was this the time? Because American financiers were still leery of railroads, the General went to Paris in 1875 to attract French investors to the project, but none were interested. Nor were Americans when Palmer tried to raise a pool of capital from ordinary, small-time investors in the Northeast.

Soon, he noticed that the fledgling Santa Fe was once again on the move, building from the Colorado border toward Pueblo—*his* Pueblo. It was alarming, but Palmer held out the hope that maybe the Santa Fe would serve his interests, feeding the Rio Grande passengers and freight despite the risk of draining his line instead, drawing away its passengers and freight onto its route to the East. Either way, one thing was abundantly clear: the Rio Grande would need to grow to compete. And for that, he needed cash.

He thought it might make sense to put off his dreams of Mexico and dive into the Rockies for a more immediate return, as he had when he reached Pueblo. He could build down just to Cuchara, seventy-five miles south, and then run west from there for a money harvest in the rich coal beds of the San Juan Valley. He might chance on a silver mine there in the foothills, too.

He knew construction would be difficult, and expensive. It would require crossing over the treacherous La Veta Pass, 9,400 feet up a steep mountainside. But the lure of immediate funds was irresistible, and he sent his chief engineer, James McMurtrie, to scope out a possible ascent. He came up with five different approaches, none of them easy. Of them, the most economical would be the most daring. It required a hairpin turn, the Mule Shoe Curve, at thirty degrees the tightest ever attempted in the Americas outside of the Andes. At midturn, passengers up at the front would be able to peer at passengers in the back as they whizzed by in the opposite direction. And even then, the climb would be so steep it would require three locomotives to haul the train up the mountainside. Once the train reached the valley beyond, McMurtrie would have to lay a tricky and expensive bridge to cross the raging Purgatoire River below.

So maybe he should go for Raton after all? Once he got over the pass, that route into New Mexico would be a breeze, as it ran forever along the flat. But, except for Santa Fe, now a thriving city of eight thousand people, New Mexico had little to offer. So it would take a long time for him to recoup his investment.

So—which route? South or west? The choice was agonizing. "So much hangs on it that it makes one's head whirl to think of the possibility of failure," the General confided to an intimate. But the bottom line was that he needed money *now*. The San Juan option offered the prospect of a quick payoff, and so he decided that the Raton Pass had to wait.

This time, the General relied on an independent concern, the Union Contract Company, to lay the track, but it proved to be harder going even than he'd feared. Before he put McMurtrie to building that Mule Shoe Curve to climb over the La Veta Pass, money ran so low, he had to cut his workers' pay and suspend interest payments on all his outstanding bonds. Thankfully, American bondholders held tight, but some Dutch investors he'd enlisted did not take the news so well. They sued Palmer in Colorado district court, demanding immediate payment. If they didn't get it, they'd force the Rio Grande into bankruptcy, and take their money from the sale of its locomotives, passenger cars, and any other assets they could recover. It was a horrifying prospect, but fortunately for Palmer, the suit went before a judge, Moses Hallett, who was himself a railroad investor. He recognized that railroads required patience and threw out the Dutch suit. The General scrounged up the money to climb over the pass and charged on into San Juan.

To Palmer's immense frustration, though, the San Juan coal did not prove nearly as plentiful, or, in that recession year, as lucrative as he'd hoped. It didn't cover the cost of hauling it out, let alone of building the track, and no silver could be found. Dangerously low on funds once more, the General had to hold off on crossing the river, for nothing much good stood close by on the far side. Only Fort Garland, once an outpost of Indian fighters, now reduced to just a few mud buildings about a public square.

But there was a shimmering possibility off in the distance. A good friend had secured a nice piece of land, the Trinchera Estate, that lay well past Fort Garland. He offered Palmer a full third of it if he could bring in a railroad. That was tempting to Palmer, since a route to Trinchera could lead on to Santa Fe, but the more he thought about it, the more he realized there was an easier way to get there—by building on south from Cuchara to go over the Raton Pass after all. Palmer sent Hunt down to Santa Fe to reassure everyone the Rio Grande was coming, one way or the other.

Ultimately, the General decided to partner up with the Union Contract Company, putting the construction costs on them in exchange for a share of the revenues. Before long, he went back to his old tricks as he built south from Cuchara toward Trinidad, high on a bluff over the Purgatoire River at the very bottom of the state. It was now a thriving near-city of twenty-five hundred people, three churches, four hotels, and two restaurants that had recently been hailed by the *Colorado Chieftain* as the state's Pittsburgh, although that may have been pushing things a little far. To become a Pittsburgh for real, Trinidad needed a train.

Palmer, however, had no intention of providing one. He wouldn't actually go into Trinidad but would stop tantalizingly short of it. Five miles short to be exact, enough that the good citizens of Trinidad could clearly see the southern terminus Palmer had planned for his line. It was just a coal field, but on it Palmer would place El Moro. This time, Palmer did not offer Trinidadians the option of paying to bring in his train as he had done to the residents of Cañon City. In his mind, the town's very success counted against it. It was too built up already for him to capitalize on a price differential, and he would do much better creating the brand-new town of El Moro nearby on land that was all his to sell. If the good people of Trinidad wanted a train, they'd have to move to El Moro to get one, the way the merchants of Pueblo had shifted to his South Pueblo. It wasn't that far, after all. If they refused to go along, the General would get by just fine shipping out El Moro coal. It was up to them. Take it or leave it. The Trinidadians left it, seething.

Meanwhile, the Santa Fe continued building west to Pueblo without any such shenanigans. When it reached Pueblo on March 7, 1876, it put its train station right in Pueblo itself, setting off an explosion of happiness when the first SF train arrived there with a blast of its steam whistle. The *Chieftain* declared the occasion produced "the biggest drunk of the present century."

Palmer tried to assure himself that the Santa Fe would now be an asset, providing him tourists to take on his Rio Grande cars up to Colorado Springs. It wouldn't suck them away from his line—would it? In Palmer's view, Strong wouldn't have to read Wellington to know that it made no sense whatsoever for a new railroad to challenge an entrenched monopoly in territory where there was scarcely enough traffic for one line, let alone two. Despite its name, the Santa Fe had to head west. It was the only thing that made any sense. So thought the General.

Still, to be on the safe side, an edgy Palmer went to Boston to let the Santa Fe board know it would not be wise for the company to veer south. The board was officially noncommittal, but the Santa Fe's president Thomas Nickerson intimated that he personally agreed with Palmer. If it was all up to him, he'd go straight on to hit the mines and coal fields in the Rockies west beyond Pueblo. Palmer found that sentiment immensely reassuring.

It would have all gone as Palmer hoped if Nickerson hadn't tired of the stresses of the western expansion and turned the matter over to William Barstow Strong.

▶◀◆◇◆◇◆◇◆◀◀

CHAPTER 6

Seeking Uncle Dick

WHILE THE MAXWELL GRANT BEGAN ON THE FAR SIDE OF the Colorado border past Raton, the valley of the Raton Pass itself was the unlikely property of Richens Lacy Wootton, nicknamed "Uncle Dick" because of his generosity in passing out whiskey at local saloons. It was to him that the two railroads would need to appeal if they sought to build over the pass. Essentially, he decided he would go with whoever got to him first—so long as he liked them well enough. Which would that be? The newcomer, Strong, or the General, whose reputation was far better established?

Wootton was a tough nut. The son of a Kentucky planter, he had ventured west on the Santa Fe Trail from Independence, Missouri, at nineteen and, after a turn as a Denver shopkeeper, established himself as a mountain man. Such a life made for many tales, not a few of them tall, which he later published in a colorful memoir.

Uncle Dick's key talent, however, was entrepreneurial. In 1852, he realized that New Mexico sheep could be sold to meat-starved gold diggers in California for a tenfold markup. He was undeterred by the fact that sixteen hundred miles of towering mountains and broiling desert stood in the way, not to mention the "savages" whose "principal business," in his account, was robbery and murder. He bought up nine thousand head of sheep in Taos, hired an escort of twenty-two shepherds, armed them all to the teeth, and set out in June. They arrived in Sacramento in October, exactly 109 days later. Of the nine thousand sheep,

"Uncle Dick" Wootton with
a friend, Jesus Silva.

Wootton still had all but the hundred he'd butchered for food along the
way. He returned home with $14,000 in gold in his saddlebags, and in
his pocket twice that in bank drafts to be drawn on banks in St. Louis.

Wootton put his money into ranching. In 1867, he saved Maxwell's
life after he was shot up in an Indian attack, and in gratitude, Maxwell ceded Wootton several thousand acres of land including the Raton
Pass. Wootton built a crude hostelry at the top and turned the crossing
into his private toll road. He let anyone chasing horse thieves pass for
free since he appreciated the endeavor, as well as Indians, figuring he'd
have a hard time getting them to pay up. But he charged everyone else
twenty-five cents a head. It was five cents more for a cow or a sheep.

Any number of dubious characters came through; for protection,
Wootton slept with two pistols under his pillow and a rifle by the bed.
Two of the more memorable visitors were the half-Cherokee renegade

Chunk Colbert and Porter Stockton, the "great rascal and desperado" who was chasing him for the reward money that had been placed on Colbert's head for murder. The two overlapped one night at Wootton's hostelry, but Colbert slipped away in the morning before Stockton discovered he'd been there. Cocked pistol in hand, Stockton had demanded that Wootton tell him where Colbert had gone. Unfazed, Wootton sent him off with a wave in the general direction he imagined the man had taken. Stockton charged off that way and ended up slaughtering a look-alike by mistake.*

The Southwest was like that in those days—lawless to the point of murdering chaos. But the Raton Pass had Uncle Dick as a presiding authority, so it was to him that Strong and Palmer would have to make their appeal. Neither man would dare stick a shovel into the Raton Pass without Uncle Dick's okay.

It was in February 1878 that Strong decided he could not wait any longer to make his move. He wouldn't build all the way down from his incoming line first. He'd leap for the spoils of Raton and build back from there. Let Palmer weave this way and that, dithering about his ultimate destination. Strong knew exactly where he was headed. Forget

---

* That left Colbert to be taken down by a renowned club-footed gunslinger named Clay Allison at an otherwise convivial dinner at the Clifton House in the nearby town of Raton, on the New Mexico side of the state line. Allison, actually, had no interest in any reward money. He'd had it in for Colbert for one of those slights that can fester on the frontier: Colbert's uncle had overcharged Allison for a ferryboat ride and then, when Allison objected, clocked him in the head. That alone didn't merit taking out Allison's ire on the nephew, but it seems to have tipped a balance that was already leaning that way. Allison sweetly made it seem to Colbert that he had no reason in the world to shoot him. He took Colbert out for a splendid afternoon of gambling on the horses. At dinner, Allison made a show of politely placing his pistol down beside his dinner plate. Not taken in by Allison's gentlemanly manners, Colbert secreted his own pistol, cocked, in his lap under the table. To get the first bullets in, he blasted his dinner companion with several rounds as soon as the entrees were cleared. Unfortunately for Colbert, the bullets were all deflected by the underside of the table. Allison plucked up his gun and deftly blew Colbert's brains out. Asked afterward why he'd risked dining with Chunk before shooting him, Allison said he didn't want to send a man to hell on an empty stomach.

Wellington, forget negotiating with towns, forget diving for immediate cash. He could see his future, and it led through Raton. His only question was when to spring for it. The answer was now.

Near the end of February, he cabled the Santa Fe board to demand financial authorization to hit Raton *now* before it was too late. He pushed so hard that, on the morning of February 26, the timid president Nickerson simply could not hold him off any longer. He authorized Strong $20,000 to proceed. To Strong, that sum was laughably deficient, but to him it spelled formal approval, and so he immediately sent a coded telegram to his chief engineer, A. A. Robinson, to seize and hold the Raton Pass. Robinson replied that he'd take the next train from Pueblo, which did not leave until seven p.m.

The very day Strong acted, Palmer acted. He'd simply gotten an eerie feeling that now was the time. He directed *his* chief engineer, James McMurtrie, to hit Raton immediately. That meant taking the seven o'clock to El Moro. He had no idea anyone from the Santa Fe might be aboard.

Perhaps inevitably, the two engineers were almost perfect reflections of their bosses, any differences strangely informing their similarities. A sturdy fellow with a robust mustache, Robinson was, like Strong, a Vermonter, and relished the chance to stick it to the snooty General Palmer. He'd worked himself up from a common laborer to chief engineer, and a tracklaying wonder. He'd directed a horde of surveyors and workmen to set down a prodigious 285 miles of Santa Fe track across the Kansas frontier in just nine months to reach the Colorado border before its federally granted land rights expired, an act that saved the company. He expected this one to do no less.

James McMurtrie of the Rio Grande was a gruff, superior Scotsman with a well-tended mustache who had fought with the General's elite Fifteenth Pennsylvania Volunteers in the Civil War, a common route for the top echelon of Palmer's officials. His technical skill was superb, and his natural smugness was justified. After all, he'd designed that astounding Mule Shoe Curve, which pushed the Rio Grande tracks into the Rockies to make the highest railroad ascent in North America. McMurtrie must have felt at home aboard one of his own Rio Grande trains now, and confident he'd gotten the jump on his rival.

Seen from the train, the Raton Pass was just a slight hollow between bumps on the horizon, barely visible in the moonlight. But before long the earth tipped up from El Moro, first gently, then steeply, to reveal the only good train route through the mountains. However speedy a railroad might be along the flat, it was not built to climb, requiring complicated switchbacks like that Mule Shoe Curve, tunnels, and more. Palmer and Strong had each dispatched surveyors—Palmer openly, Strong in secret—to determine how it might be ascended, and then descended on the far side without flying off the rails. Now it remained for McMurtrie or Robinson to enact them, whoever got there first.

When the Rio Grande train finally pulled into the station shortly before midnight, the two chief engineers disembarked onto the railway platform furtively. Both of them were eager to meet with Uncle Dick, but it was pitch dark, and the train ride had been tiring. One of them checked into the town's only hotel for some rest, figuring an approach to Uncle Dick could wait until morning when such a delicate conversation would surely go better over coffee.

CHAPTER 7

Santa Fe, Inc.

PALMER MAY HAVE DIRECTED THE RIO GRANDE LIKE A personal enterprise, but the Santa Fe pretty much ran on its own. That was the nature of a corporation. Strong provided the guidance that came from his Grow or Die philosophy, but he lent it none of his personality, and certainly none of that "lofty plane of thought and purpose" the Rockies inspired in Palmer. Unlike Palmer, he had nothing to do with his railroad's founding. He was just twenty-two and still with the Milwaukee & Mississippi in 1859 when the Atchison, Topeka and Santa Fe came into being in Kansas.

It was the brainchild of Colonel Cyrus K. Holliday, a stout, earnest, loud-talking entrepreneur whose full beard gradually evolved into a mustache that curved up into his sideburns, creating the impression of a permanent, oversized smile. Like Palmer, he had grown up in Pennsylvania and entered the train business as the smart career choice for any enterprising young man. But there the similarities ended. Colonel Holliday never actually got into the Civil War; his rank was largely an honorific, and he had a college degree to get him into the law. He did his early railroading from the legal angle, working for a railroad construction company that went bust—but not before he'd made $20,000 to invest in Kansas.

Kansas had been conceived as a vast reservation for the Indians

A full-bearded Cyrus K. Holliday
in his early Kansas days.

who were being forcibly removed from the East.* When white settlers wanted that land, too, the Indians were removed once more, to
Oklahoma. This was 1854. By then, slavery had emerged as a threat
to the republic. While the famous Missouri Compromise of 1820 had
kept the number of free states and slave states in political balance,
Kansas posed a huge question. If it joined the Union, would it allow
slavery or not? To keep the nation's fragile peace, Congress passed
the Kansas-Nebraska Act which let Kansans decide the answer for
themselves through a popular vote. But that decision unleashed a pan-
demonium, as partisans on each side of the slavery divide flooded in
to win the vote, creating a violent uproar that turned the future state
into "Bleeding Kansas."

Cyrus K. Holliday was one of the newcomers. He arrived as an abolitionist, not because he had strong feelings on the issue, but because he

---

* The territory itself had been named for one of the largest of them, the Kansa, or "People of the South Wind," a term that describes all too well these native peoples who'd
been blown about by ill fortune.

expected it to be the winning side and hoped to capitalize commercially on his insider status. He came by steamer down the Missouri River a few months after the act was signed. By then, he'd married a dairyman's daughter. Since his wife was pregnant, Holliday had thought it wiser to leave her behind in Pennsylvania until Kansas calmed down. He thought that would just take a few months. It took five years.

While neighboring Missouri was well settled, Kansas was wide-open prairie, and Holliday found it glorious. "God might have made a better country than Kansas," he wrote his wife, "but so far as my knowledge extends he never did." With some other newcomers, he scouted a place to establish a town of their own. They settled on a strip of land along the Kaw River called Topeka—an Indian word meaning a "good place for potatoes." It stood at the junction of the two great trails into the west, the Independence Trail to California, and the Santa Fe Trail to New Mexico. To him, that was inspiring.

Winter was coming on, and the group had to sleep in tents on the frozen ground. To keep warm, Holliday wore all his clothes, even his hat, and wrapped himself in two blankets and a buffalo hide. The men spent their days marking off sixty-two acres for a town. Holliday saw it as the future capital of Kansas, a vision all the others found hilarious.

Kansas was not exactly hospitable. That first winter, temperatures dropped to thirty below, and the summer brought terrifying storms whipping across the plain. "When I commenced this letter I said there was a thunder storm raging," Holliday wrote his wife. "It is now *raining, hailing, blowing, thundering & lightning* all at the same time." Malaria and cholera were rife, and it was not uncommon for a Kansan to wake up to find a rattlesnake coiled under the bed or, worse, draped from a bed pole overhead.

Nonetheless, Holliday organized a land company to purchase title to the sixty-two acres they plotted, and he bought thirty of the house lots himself. He was chosen the company's first president and became an agent for the Massachusetts Emigrant Aid Society, luring antislavery voters from the East. He also helped organize a "Free-Soil" antislavery party and ran as its candidate for the territorial legislature. In this, he fell short, since proslavery voters, most of them from neighboring Missouri, still predominated.

Holliday could see that the antislavery tide was rising, but the problem was, as it did, violence was breaking out with the proslavers. Here a lynching, there a murder, another place a mob riot, and in still another a pitched battle that brought thousands into the fray. "Bleeding Kansas" indeed.

Undeterred, Holliday organized the all-volunteer Second Kansas Regiment, making himself its colonel to take on the proslavery "Border Ruffians" surging in from Missouri. In 1859, he became a delegate to the convention to determine the terms of statehood and managed to make little Topeka, population seven hundred, the state capital after all, and himself its first mayor. It was finally time to bring his wife and daughter to join him.

Holliday had initially thought he'd invest in steamboats, but now he decided to go with a train. He was thinking of just a short line from Topeka to Atchison, fifty miles to the northeast. Kansas railroads were the rage there at the exact center of the country, the birthplace of two dozen railroad companies. Following the example of the Pacific Railroad, Holliday figured he could get the federal government to pay him to do it. After he won a state charter for his railroad, he wrote out a federal bill to do just that. By now he'd packed his board with Kansas dignitaries, and in 1863 the Congress voted in his bill without changing a comma. It gave Atchison and Topeka three million acres of Kansas's public land.

The bill required the line to be completed across the state in ten years, which seemed like plenty of time. But the war interfered, construction money dried up, and a series of droughts and cyclones created havoc. Five years in, not a single rail of Atchison and Topeka track had been laid, and the company was broke. But Holliday discovered that some Pottawatomie Indian land had been awarded to another railroad line that had been unable to use it. Holliday snapped it up on the cheap, and then sold it for full value to get enough cash to send his line seven miles from Topeka toward Atchison. That wasn't much, but Holliday figured it called for a celebration. He sent a trainload of distinguished Kansans at the hair-raising speed of fifteen miles an hour and transferred them to wagons to reach the little town of Wakarusa, where Holliday delivered himself of an epic stem-winder. In a fit of oratory, he declared that his train would not stop at the end of Kansas. Dramatically throwing his arms over his head, then crossing them at the elbows, his fingers outstretched to heaven, Holliday declared that Topeka would be the starting point for Kansas railroads going west, just as it had been for the pioneers before them. Trains would come to Topeka from Chicago and St. Louis, as represented by his two up-thrust hands. And they'd travel down his arms to enter the Southwest at his shoulders and run down his body to the Gulf of Mexico and California at his feet. "The coming tides of immigration will flow along these lines of railway," he thundered, "and like an ocean wave to advance up the sides of the Rockies and dash their foaming crests down upon the Pacific slope!"

The local newspaper thought it would be enough just to get to Emporia, sixty miles southwest, but Holliday would not be deterred. He added Santa Fe to the train's name, and then, when the newly christened Atchison, Topeka and Santa Fe Railroad had gone just seven miles more, he brought in another crew of dignitaries to celebrate again. Now that the line had finally shown some progress, Emporia bought enough bonds to receive Holliday's train two years later. From there, Holliday ran tracks to Newton (so called because it was indeed a new town) and then to Wichita. Then, with A. A. Robinson directing the tracklaying, double-time on due west to the Colorado border, to beat the federal deadline by three months. By then, Holliday had surrendered the title of president to Kansas's first senator, who could do the railroad more good, staying on as just one of many directors. It no longer needed a visionary. It needed moneymen from the great capital centers of Philadelphia, New York, and Boston.

Of the three, Boston was the most railroad minded. It had been the nation's first major hub, with any number of lines running in and out of the city. For a small city on a stub of a peninsula, Boston was surprisingly worldly. The first great Boston fortunes were made in the China trade, then amplified by investments in the textile mills of early industrialization, only to be compounded by provident marriages to the daughters of other wealthy Bostonians.

The investors were referred to as the Boston Crowd because of this interbreeding. They were mostly drawn from that class of Bostonians—Brahmins was the derisive term—whose members were, in the main, Harvard-educated Episcopalians, with the occasional Unitarian thrown in, who lived for their clubs and thought of themselves as existing on a social plane only slightly down from God. They were like honeybees in a hive—industrious but interchangeable.

From this esteemed collective about the only one to emerge with a distinctive personality was Thomas Jefferson Coolidge—and he on the strength of a surprisingly frolicsome memoir he had privately printed, the copies limited to just forty-eight—who served as the Santa Fe president for a year starting in 1880. Coolidge took his middle name from the American president, his great-grandfather on his mother's side, but his lineage could be traced back to a Coolidge who settled in Watertown, just downriver from Boston proper, in 1630.

Coolidge grew up in Canton, China, where his father was a partner in a Boston trading firm that dealt tea and opium in the China Trade. After Harvard, he married the daughter of William Appleton, clipper ship owner, European trader, president of Boston's Second National Bank, congressman, president of the Massachusetts Hospital, and one of the richest men in the city.

Coolidge served as the ambassador to France, and on any number of important national commissions, but he prized above all else his membership in The Friday Club, which, he noted, had once blackballed the eminent Dr. Oliver Wendell Holmes, father of the Supreme Court Justice, for fear he would dominate the conversation. His memoir is rich in frivolity—meeting the actress Fanny Kemble at a monastery in the Alps, touring the Caribbean aboard his eighty-foot yacht, riding a donkey to the temple of Ramses in Egypt.

In recounting his life, Coolidge mentioned his yearlong presidency of the Santa Fe railroad only in passing. In October of 1878, he wrote, he took his son and namesake for an extensive tour of the West that covered almost ten thousand miles, particularly enjoying the Colorado portion aboard the Santa Fe Railroad with one William Barstow Strong. Coolidge did not mention that he was a director of the company at the time. When he returned to the West two years later, he noted that he again boarded a Santa Fe train, but this time stepped off at Topeka, where "I was elected president of the Atchison Railroad at the annual meeting."

Six sentences later, he was done with it. "I resigned as soon as I could," he wrote. "I think in about a year and a half." He failed to mention that on assuming the presidency, he had purchased $700,000 worth of the company. The point being, he presided over the Santa Fe only as an investor. The moment the investment seemed unpromising, he sold it and got out.

None of the Boston Crowd served as presidents for more than a few years. But this was the way of the modern corporation as they came to define it. They were not managers, but investors, and that inclined them toward caution in a business that demanded daring. It was probably inevitable that the Boston Crowd's enthusiasm for the presidency of the railroad operation would diminish as the company ran farther into the unruly West. It was one thing for the Santa Fe to serve eastern Kansas towns like Atchison, Topeka, and Emporia, all of them fairly civilized places populated largely by the New England emigrants who'd come to save Kansas from slavery. Much past that, the social distance from Boston became too great, and the gentlemen of the Boston Crowd ceded operational control to professional managers.

An early point of tension came in 1873 when the Santa Fe sought to market some cropland north of Wichita in the middle of the state. In some desperation they had looked clear across the ocean and then some ways into Europe to recruit thousands of Mennonites—a dark-clothed Amish-like people who were being persecuted for their religious practices by Czar Alexander. Learning of their plight, and hence their possible availability, the Santa Fe invited a group of Mennonite bishops to cross the ocean at the railroad's expense to look over the land the Santa Fe had for sale. When they approved, the Santa Fe sent a German-speaking grocer named Schmidt to Russia as its emissary to visit fifty-six Mennonite villages around Alexanderwohl, Russia, by sleigh, ultimately to recruit four hundred families willing to buy 160-acre spreads at five dollars an acre. One of the bishops who'd originally approved of the deal brought his family over with him to resettle on the Kansas plain. When he arrived, he looked around at the desolate, windswept landscape, sat down, and wept. He wrote:

> I thought of the poor families with their children. We had no provisions, no friend in the new world, the winter was nigh at the door, we were wanting of dwellings, provisions, agricultural implements and seed; everything was high in price, some of our people were old, weak and sick, the future seemed very gloomy; there were also no prospects of rain, only windy, dusty and very hot, all of this fell over me, so I could not help myself but leave my tears free flow.

But a Kansas neighbor assured him that the winter would pass, the rains would come, and the crops would rise. Sure enough, he put up a sod house on his parcel, plowed the field, planted wheat, and learned, as he said, the "dear Heavenly Father has a watchful eye" over him. They prospered there, as did the Santa Fe, and Kansas, too. For they brought with them a hardy, cold-resistant variety of wheat, Turkey Red, that could survive Kansas's brutal winters, unlike the local strains then in use. Before long, Turkey Red made Kansas the greatest wheat-producing state in the country, generating a fifth of the national harvest.

The final break with the Boston Crowd came later at Dodge City. If any town in America was going to be too much for them, that was it. By the time the Santa Fe was venturing across Kansas, the buffalo had largely disappeared and Dodge City was a cow town, scooping up the cattle to ship them to slaughterhouses in Chicago and St. Louis before the herds roamed farther north to be taken by the Kansas Pacific.

In claiming the town for the railroad, the Santa Fe laid its tracks right down the middle of it, creating "right" and "wrong" sides that were even more pronounced than usual. In Dodge, the wrong side offered enough gambling, drunkenness, lewd behavior, and criminality to make Dodge the "wickedest" city in the west. There were fifteen murders in Dodge City's first year alone, most of them shockingly casual. A typical victim was the black hackney driver who was shot in the head by a drunken cowboy who stole his carriage. The corpse was left sprawled in the street, where some passersby shrouded it under a buffalo hide. When the killer came back to peel off the hide and examine his handiwork, he proudly pointed to the spot his bullet had entered the dead man's skull. The body was dumped in the town cemetery without a coffin. No charges were filed.*

Appalled by all the mayhem, the Santa Fe railroad threatened to shut down the cattle business if the cowboys did not mend their ways. But the cowboys kept on drinking and brawling and shooting off their guns just as before. Santa Fe officials asked that the saloons please at least observe the Sabbath, to no avail. The cowboys shot out the headlights of the incoming locomotives and fired off pot shots at train crews.

Initially, law enforcement was left to a sole town marshal. To beef it up, the railroad authorized the marshal to add a deputy in 1876. He selected the taut, even-tempered Wyatt Earp. Like most lawmen in the period, Earp had started out as a lawbreaker, getting himself arrested for a saloon scuffle in Wichita. When some louts kept on fighting him as he was taken to jail, he told them to knock it off in a tone that somehow got them to stop. Taking notice, the Wichita marshal thought that maybe Earp would make a decent deputy. After he arrived in Dodge City, Earp was allowed to hire a fellow deputy. He signed up his friend Bat Masterson, a level-headed character who favored a derby hat. Mas-

---

* African Americans, it should be said, made up a disproportionate share of Dodge's homicide victims, and demonstrated the mindlessness of the slaughter. Black lives simply weren't worth very much to a drunken cowboy who had likely fought for the Confederacy, a bitter legacy, perhaps, of the state's fight over slavery.

Bat Masterson in his trademark derby hat.

terson had fought Indians as a Texas Ranger, but, in a typical saloon mishap, he'd been struck in the hip by a stray bullet shot by a drunken sergeant who was losing a card game. Tumbling to the floor, Masterson blasted the sergeant in the chest with his Colt. Afterward, he walked with a limp, but his disposition was unaffected and he remained a very good man with a gun.*

The Santa Fe decided to revise the pay structure to encourage the arrest of suspects rather than just shooting them on the spot, and so it offered Earp and Masterson $2.50 per arrest, instructing them to shoot only to maim in order to preserve the suspects for trial. Once the two deputies started in, they made three hundred arrests a month between them.

---

* Masterson had a way with words, too, ultimately working as a newspaper columnist in New York City when he died in 1921.

It was a lot to handle, and Earp and Masterson soon hired Masterson's brother Ed as a third deputy. When a crew of rowdy Texas cowboys started shooting up the Dodge City dance hall, Bat was first on the scene, Ed hurrying in after. Ed encountered a drunken cowboy inside the door, and politely asked him to hand over his gun. The cowboy shot him in the gut instead. Ed stumbled out into the street, his jacket smoldering from the gun blast. Bat shot his assailant, then rushed to his brother, who'd fallen to the ground by the railroad tracks. Bat pulled his brother up to comfort him, but Ed died in his arms.

By then, Thomas Nickerson had assumed the presidency of the Santa Fe. Typical of the Boston Crowd, his brother was also a director, and his son would soon be, but Nickerson was a cautious man, and this kind of frontier ruckus was too much for him. The Santa Fe had previously hired a superintendent, T. J. Peter, to oversee the building through Kansas, and, unlike the good gentlemen from Boston, Peter filled the Kansas sky with his profanity if things weren't done to his liking. In 1877, Nickerson realized he needed another Peter to handle the West for him. He found his man in a rising star with the Chicago, Burlington and Quincy.

CHAPTER 8

▸┄●┄○┄●┤┄●┄○┄●┄◂

Enter the Queen

IN JANUARY 1869, THE GENERAL WAS ABOARD A PENN-
sylvania train rattling through the farmland of southern Ohio in a
drenching rain when he met the woman who would upend his life.
He was still with the Kansas Pacific, but, fresh from his investigations
of the Southwest, he was starting to think about creating a railroad of
his own. He was taking in consideration of the financial requirements
when he fell into conversation with a somewhat wizened, middle-aged
gentleman in his compartment. The man was William Proctor Mellen,
who'd served under Secretary of the Treasury Salmon P. Chase in the
Lincoln administration. He was a New York lawyer now, but he happened to mention he'd collected some wealthy British friends who'd
taken an interest in investing in the American West. So Palmer was
already warming to Mellen when his daughter quietly entered their
compartment to settle herself cozily beside him.

Mary Lincoln Mellen was just nineteen, a bewitching, Pre-
Raphaelite figure with a low, musical voice, darting eyes, and wild hair
that ran rapturously down over her shoulders. She had been nicknamed
"Queen" by her grandmother for her amusingly royal ways as a child,
and the name stuck, for it captured something about her independent
nature. In the compartment, Palmer likely said little to her beyond the
limited pleasantries a gentleman exchanges with an attractive young
woman in the company of her father, but he clearly found her intoxi-
cating. After parting with her at Cincinnati, Palmer pursued her with

The wild-haired Queen Mellen
when she first met Palmer.

letters and just two weeks later he plunged into the maelstrom and proposed. No less impetuously Queen accepted, even though she had not
laid eyes on this impassioned General since the train ride.

Queen had been born in Prestonburg, Kentucky, home of the vicious Tenth Kentucky Cavalry that had pestered Palmer's Fifteen
Pennsylvania in the war. But now, Palmer's only thought was that the
place gave Queenie's voice a lovely, Southern lilt. Her mother had died
of a "brain fever" when Queen was just four, leaving her to be raised by
her maternal grandmother, the one who nicknamed her, until she left
home to attend the Cincinnati Institute for Young Ladies. By then her
father was off in Washington working in the Treasury Department, a
position from which he ultimately resigned after some underlings were
accused of taking bribes to manipulate the price of federal cotton during the war. Although Mellen had not been implicated, he felt he could
not continue after his honor had been impugned, and so he moved
with Queen to a big house in Flushing out on Long Island.

It needed to be big because he'd remarried by then—to his late wife's sister, Ellen, twenty-two years his junior, and he'd had six more children by her. If it was bewildering for Queen to find herself awash in half-siblings from a stepmother who doubled as her aunt, she hid it well. She grew into a fiercely independent young woman, with a passion for art and music and an eagerness for friendship that was, as she wrote in a school essay, "unlimited and unreserved." She also held a host of feminist convictions about the possibilities of womanhood. A woman, she argued, shouldn't be thought "useless or wretched" if she chose not to be a mother. Palmer, however, wanted her to be just that, and to yield to a headstrong general who was bent on bringing her west. Palmer offered wealth, adventure, cultural refinement—and a way out of a life that must have felt crowded. To all of this she could only say yes, but in her heart she was anything but certain. In those early, ardent letters of his, Palmer called her "Queenie." But she called him "General" long after he let her know he preferred she call him "Will."

If Palmer had wooed Queen impulsively, he showed restraint once he'd secured her. All part, perhaps, of his trademark push-pull as a Quaker warrior. Once she'd agreed to the marriage, Palmer didn't go see her for another two months, and when he did so, he announced himself at her Flushing home by leaving her his formal calling card as if he were a near stranger. "To Miss Queen Mellen with the kind regard of William J. Palmer," went an early one. "May I have the pleasure of your company this evening to Gen. Kirkpatrick's lecture—on Sherman's March to the Sea." Dozens more such invites to refined cultural events followed, but his letters to her gradually edged into a soulful intimacy with this wild beauty half his age, as if she had unlocked something deep within him.

It started when he thanked her for loaning him a pair of popular occult romances by Friedrich de la Motte Fouqué. The novels were *Undine* and *Sintram and His Companions*, each named for its main character (the first female, the second male), who had come, possibly, to serve as a stand-in for her and him. *Undine* was a cheery, springtime story of a water nymph who marries a knight to gain a soul, *Sintram* a dark winter's tale of a guilt-ridden lad desperate to purify himself to secure his mother's love. As a fallen Quaker, Palmer may have seen too much of himself in the book's hero. "I fear you will think I am so bad," he wrote Queenie sheepishly, "that I ought to read [that book] over again."

A passing remark, one might imagine, but he returned to the theme of guilt in another note a few days later. He'd been brooding about it, he wrote her. He'd been sick for several days, and lying in bed had given him a chance for "a serious thinking over." A remarkable outpouring followed, a kind of cri de coeur from a man who'd never before had anyone to spill his secrets to.

> I do not believe much in confessions or promises of amendment until the result is achieved—so I will simply state the conclusions that I came to—and these very briefly—1st. That the wickedest man in New York was nothing to me. 2nd. That I intend to be good. My life has been a chequered one, with all sorts of experiences among all sorts of People—I cannot say it has been an unhappy one, because I have been too reckless to care for consequences, or the opinions of the people. When I went to the War, I never expected to come back, and when having seen better men fall everywhere around me, I came out without a scratch. I carried into civil life a good deal of the fatalism of soldier. But my creed has always been better than my practice—and hereafter I am determined that they shall be made nearer to square—not by lowering the former but by bringing up the later—so that with your love I expect to attain, even in this life a share of that positive happiness which is so different a thing from the absence of misery, or from stoicism.

It could well be that Palmer felt nothing more than survivor guilt for living through a cataclysmic war that had claimed so many other, worthier men—or, perhaps, mortification for failing to live up to his Quaker ideals by joining the cause in the first place. Was this just the standard confessional of the pious, an expression of humility, and no more? Or was this determination to be good alluding to something specific that he may have blurted out to Queenie in his fevered courtship?

If so, what? Other than Queenie, there were few others he spoke to about such things. He'd always been one for close male friends, of whom the most recent was his tentmate Willie Bell—but one of his confidants was a fellow Quaker boy, Isaac Clothier, to whom he had written so many warm personal letters, Clothier bound them into a private volume after Palmer died to remember him by. While Palmer inquired many times about his friend's blossoming romances, he confided none of his own, leaving Clothier to conclude there weren't any. Interestingly, the correspondence stopped in July 1868, just six months before he met Queen, as if she took over from his boyhood friend. Conceivably, at some point, Palmer may have engaged in a deeper level of intimacy—with Clothier or some other inviting young man—that he hadn't felt right about afterward, but it is impossible to know. It's also hard to tell how much it really mattered. The word "homosexual" did not exist in common English until 1892, when it appeared in a translation of a German tract on sexual practices. For some time, the crude term "sodomy" had been used to describe a set of furtive, same-sex acts that were considered abominable, but the very concept of same-sex relationships built on mutual and natural affection seems not yet to have come into being. In the Civil War, the acts were forbidden as "crimes against nature," but tellingly no soldier on either side was court-martialed for committing any of them.

Whatever transgression Palmer was referring to, if transgression it was, neither he nor Queen mentioned it again, although the question about Palmer's sexual preference would return. Whatever the reason, he seemed to be grappling with an inner torment not so different from what afflicted him in the run-up to the war. In his blossoming romance with Queen, he was once again caught between wanting what he did not want, and not wanting what he did.

Most of Queen's letters back to him have been lost—it is revealing that he was not more careful about keeping them—but by all accounts it seems that Queen was not troubled by anything that he might have done. She was bothered by his unavailability, repeatedly vexed by how Palmer's railroad commanded so much of his attention. (Clothier, curiously, had made a similar complaint in a letter in 1859—claiming that the "idol" he worshipped was certainly not female, but rather "the latest locomotive in which your soul has seen written perfection.") The spring after he proposed she teased him: "[I] hope to see you sometime before the Christmas holidays—if the railroads and tunnels can share you!" before giving in to pouting. "You can't be very sad at the thought of your forsaken sweetheart, when your beloved 'wife' is with you; I suppose she has hardly left your arms since you left me!" The patience and lightheartedness quickly faded when she realized his priorities were not shifting in her favor. "How pleasant it must be to enjoy her society without any Queen to interfere with her complete monopoly of you— Sweet thing! let us fervently hope she will never 'wear out,' that her love will always be as warm as it has been hitherto . . . Three cheers for such a faithful 'wife.'"

Palmer did his best to assure his fiancée she was "unrivalled" by any train, and Queen did her best to resign herself to the situation. "Where duty interferes with pleasure, pleasure must always give way," she wrote him, "and I shall always try to make it easy, instead of difficult as I have done or two or three occasions." Still, off to Chicago on yet more railroad business a few weeks later, Palmer couldn't help feeling that, in decrying his devotion to his railroad, Queenie was abandoning him. "Do not hesitate, my darling, to tell me how much you love me," he begged her. "You cannot spoil me in that way. You would not fear to if you know what an inspiration it is to me—and not only an inspiration, but a shield."

In spite of their efforts, Palmer's railroad did separate them. It drew him west while Queen stayed in the East, not just by circumstance but also by temperament. To her, the vast western landscape that so enthralled him was an alien, forbidding and, worst of all, dull territory without any museums, opera houses, and salons. Palmer could not have disagreed more, and he pleaded with her, time after time, to reconsider. "Is it any wonder that—even when the Princess is left out—young men, taking the circumstances for the reality, should fall in love with a life like this, and riding year after year in this free air across these lonely plains." Surely, he thought, there was something in the West that would thrill her, too, and if he wanted to make their marriage a success, he'd have to find it.

‣━◦━◦━◦━‹

Six months after meeting Queen, Palmer found himself atop a stage-coach gazing at the majestic Pikes Peak. He'd climbed up with his bed-roll to sleep under the July night sky and as the sun rose to reveal the mountains, he was in raptures, as if he was glimpsing the divine. "I could not sleep any more with all the splendid panorama of mountains gradually unrolling itself," he wrote Queenie. "I sat up and drank in, along with the purest mountain air, the full exhilaration of that early mountain ride." He bathed in the chilly waters of a meandering stream, then, eager to see more, he hiked up into the grassy canyon, fringed with spruces and sprinkled with lakes, that led up to that wondrous, boulder-strewn Garden of the Gods where he would soon place his cas-tle. Surely, he told himself as he looked about the craggy, snow-capped peaks rising off the plain, this vista would lure her west. Before the week was up, he enlisted Alexander Hunt to secure several thousand acres for a future town he'd call Colorado Springs, although in fact the nearest springs were in neighboring Manitou. He was convinced that Queenie would love it and started laying out a town for her filled with squared-off streets named for local streams like Kiowa, Bijou, and Platte, rather than the usual letters or numbers. He planned to line them with shade trees to evoke the nicer towns of the East Queenie was used to, and he'd make the lots unusually large, a full 50 by 190 feet, with the expectation that the houses built on them would be grandly commensurate. He left plenty of room for gardened parks, too. Also churches, schools, music halls, theaters, and museums. All of it exactly what she would want.

By mail, he spun Queenie his vision of the town, and the great house he'd build for her in it. He'd place it under the "bold, pine-topped hills near the mountains" on a spot of her choosing, although he had a few in mind himself. A "Dunkeld," he called it, after a Scottish palace, rich in the Ossian legends she loved, "the child of our fancy and creation."

They'd have a deer park for antelope and black-tailed deer, a range for the last of the buffalo, and a preserve for peaceful Indians. And Queenie would have society, too, as there was plenty of room for homes for their friends—"those really our friends," he emphasized. They'd surely come to a place of such "health and vigor."

Soon, he didn't think of Colorado Springs as a town at all. It would be a club, literally, called the Fountain Colony. Memberships would be available only to those persons the General judged to be "of good character and of strict temperance habits"—and all of them to Queen's taste, or course.*

There together amid such a splendor of friends, views, culture, and social refinement, he rhapsodized, "life would be poetry—an idyll of blue sky, clear intense atmosphere, fantastic rock, dancing water, green meadow, high mountain, rugged canyon, and distant view of the kind that gives wing to the imagination and allows no foothold for it to halt upon short of infinity."

He was sure that it would take just one look, and she'd be smitten.

---

* The idea of turning a town into a private club was just a more benign version of a more drastic development that was occurring in the industrial cities of the East. As class tensions rose, private military fortresses were rising up to protect the neighboring elite. Armories was the term for these massive, stone, medieval-style redoubts. But they were not intended to repel foreign invaders. They were meant to save the wealthy from a revolutionary mob. Each armory provided a vast indoor space for the training of a private army; stout walls and triple-thick doors to withstand dynamite and battering rams of insurgents; and rooftop crenellations to shield riflemen firing down on any revolutionaries massed on the street below. One of the grandest of these many fortresses went up on New York City's Park Avenue, between Sixty-Eighth and Sixty-Ninth Streets, in 1877 shortly after the Great Railroad Strike. Today, it is thought of only as a genteel if oversize palace for art and music. But it was intended to give the wealthy a place to ride out a revolution in comfort. Inside, besides a shed for the drilling of the elite volunteers who were dubbed the "Silk Stocking regiment," the Park Avenue armory featured private rooms that would not have been out of place in the most sumptuous social clubs in the city. There was a plush Veteran's Room designed by Louis Comfort Tiffany in a mélange of Persian, Japanese, and Celtic motifs; a barrel-vaulted library featuring an elegant ten-foot chandelier of Stanford White; and the showpiece, a Renaissance Revival staircase created by the eminent George C. Flint and Company, lit by bronze torchieres. Such social anxieties added to the desires of the wealthy to seek refuge in places like Colorado Springs.

That fall, the General invited his old friend Willie Bell to visit the property. Finding him "the same good old fellow he always is," Palmer had
the most glorious time with him riding horses all about. It was as if
something had been released within him. When they came across a
herd of wild horses, they gave chase just for the fun of it, rousing all the
other wildlife—antelope, prairie dogs—as the two of them thundered
past. "We flew so fast," Palmer later recounted to Queenie, "that I could
scarcely see even the other wild things that started up from every little
cove and grassy slope, alarmed by this sudden invasion of their quiet
sanctums." Bell was so captivated, he made plans to move there himself.

In January 1870, a line coming down from the Union Pacific at
Cheyenne was finally linked to Denver, and it would soon be followed
by the Kansas Pacific coming from the east. It was time for Palmer to
put together a railroad of his own to take advantage of this new hub.
The vision of his railroad had come to him in a waking dream earlier
that month while he was staring out into the blur from the window of
a KP railroad car rolling through Salina, Kansas. Full of excitement, he
immediately described it all to Queenie:

His railroad wouldn't be any sort of rapacious cutthroat operation,
he assured her, but a family—intimate, happy, loving. A "little railroad,"
as he put it, of just a few hundred miles, all of it "under one's own control with one's friends, to have no jealousies and contests and different
policies, but to be able to carry out unimpeded and harmoniously one's
views in regard to what ought and ought not to be done." In a word,
"Quakerish." Not only would he put up schools, bathhouses, libraries,
and educational programs for the workers, he'd give them shares in the
railroad, too, so there would "never be any strikes or hard feelings."

By the end of the month, he told Queenie he wanted her to move
to Colorado Springs with him. In his mind, it was all set. Surely, she'd
see this little family of his as *her* little family, too. "I felt so happy that
I would have such a wife, who was broad enough, earnest enough and
good and pure enough to think that a wild home amidst such scenery

were preferable to a brown stone palace in a fashionable city; to go out each evening on some neighboring hill and find each time a new vision of beauty and grandeur!" She must not have shared his enthusiasm, however, because he promised to "talk it over fully" with her when he next saw her. Although over a year had passed since she had accepted his marriage proposal, a date for the wedding had still not been set. Nonetheless, the General charged ahead with plans to bring a train from Denver down to Colorado Springs, and to develop Colorado Springs itself. The train would not come roaring by their castle, he promised her, only pass by near enough to look "graceful" gliding cross the valley from their windows. "It won't hurt—when it is our own railroad, will it?" he asked plaintively.

⊷⊶⊙⊷⊶

It wasn't until April that Queen finally came west to see this utopia for herself. She rode west on the Kansas Pacific with her father and a small group of potential investors to Kit Carson, a small outpost fifty miles from Pikes Peak, where the General had arranged for coaches to receive them. As the coaches jerked along on the uneven road, the view out the window to the barren snow-swept plains of eastern Colorado could not have been less welcoming. Where was the Colorado Springs of her Will's dreams, Queen must have wondered? Where were the theaters? The clubs? The fine houses? All she could see were a couple of dreary log cabins, darkened by woodsmoke on the interior. The stunning peaks that the General had gone on about were hidden behind dark, threatening clouds that had settled over the valley and would not lift. And it was so cold! It may have been spring, but it felt like February, the air miserably cold and wet, encasing her in ice. When Queen finally caught up with her husband-to-be, she put on a show of enthusiasm for him, but her heart was not in it.

One morning, the General took her riding out to see the perfect site for their "Bijou," as he termed their glorious castle-to-be. It stood above Queen's Canyon, which he'd named for her. In summer, the canyon was an enchanting jumble of various smooth, ruddy sandstone, a spill

Colorado Springs as it looked to Queen Palmer,
once the snow had cleared, after she arrived in 1870.

of massive reddish jewels—or so Palmer said. Now, as she looked about, all she saw was a dreary vale of icy snow.

The worst realization, however, came when a local took Queenie aside to show her the exact spot where a couple of boys had been slaughtered by some marauding Arapaho two years before. He thought she'd be interested. She was appalled. Her uncle Malcolm Clarke, a Montana rancher, had been axed to death by a suddenly enraged Piegan Blackfeet Indian the previous summer over what was meant to be an amicable dinner. In furious retaliation, the US Army had mowed down two hundred Piegans with Gatling guns, but Queen was hardly reassured. To allay her fears, the General enlisted private soldiers to protect them from any Indians who might do them harm, but just the fact that he'd had to do that made Queenie terribly uneasy.

To distract her, Palmer brought her to see the rock formations of the Garden of the Gods, where he laid out his architectural vision for their Bijou. She said she'd like to call it Glen Eyrie. Warming to the prospect of a home as she explored the site, she suggested it have four hexagonal rooms like the cells of a beehive, all clustered about a big stone chimney, providing fireplaces throughout, with high windows offering grand views. Together, she and Palmer plotted out which rooms would belong to the children—a boy and a girl, they hoped—and which would be given to a teacher to educate them. Palmer must have found it an immense relief—until a young Indian chief came around to the cabin, excited to learn that a beautiful, unmarried twenty-year-old woman was in residence. He came brightly gartered for courtship and, trailed by four handsome mustangs, he planned to offer for her hand in marriage. Unamused, Queen sent him packing.

When Queen returned to Flushing, her mind was in a swirl. Could she really live in Colorado Springs? It had possibilities, she had to admit, but it was so dreary—and scary, too. There were wild animals, untamed wilderness, unpredictable elements, and of course the threat of hostile Indians. Palmer assured her that he himself always went about "well armed" to protect himself, but she had to wonder: What about her? Was she supposed to strap on a brace of pistols every time she went outside?

A decision loomed, for that summer the General completed the Kansas Pacific line to Denver, freeing him to start his own railroad. Relying on investors from the Maxwell Grant syndicate, plus some other railroad men in Denver, and a few eastern financiers like Queen's father, the General formally incorporated his Denver & Rio Grande Railway. He installed himself as president, made Willie Bell the president of the construction arm, and placed his future father-in-law on the board with a number of Colorado luminaries. He was ready to build down from Denver to Colorado Springs.

But first he had to rush to New York. There, he would talk up the venture in the business press and solicit further investment—and marry his Queen, the capstone to the great arch of his triumph. Queen, however, was still rattled by her visit to Colorado Springs. Her one surviving letter from this period reveals that she admitted to Palmer, whom she called "my Will," that she was "very, very tired." She tried to assure him she loved him with all her heart, but her anxieties came through as she contemplated the future. "Three weeks from tonight, where will we be?"

Despite her anxieties, the wedding was held on November 8, 1870, exactly two weeks after Palmer's railroad was formally launched. The service was performed at Queen's father's house in Flushing and, according to the *Flushing Journal*, only "a few friends" attended. The bride's gold wedding band was simply engraved "WJP to Queen" with the wedding date. The whole event had a rushed, haphazard quality, as if it were a mere technicality.

The next day, the married couple boarded the steamer SS *Scotia*, bound for a three-month honeymoon of European tourist sites, with plenty of time set aside for the General to see potential investors and study the latest developments in European railroading. Even as Mrs. Palmer, Queen saw a rival in her husband's train. After they arrived, Queen was obliged to tour her beloved museums alone, but she was pleased by the European high life and recorded all her impressions in her journal in great detail. The final entry recounts a visit with Palmer to the family of the Rev. Charles Kingsley, an illustrious man of letters whose daughter, the cheery and intrepid Rose Kingsley, immediately won Queen's heart. She'd gone to bed that last night "wishing that we were to spend weeks instead of days with our new friends." In the morning, the Palmers began their journey back on the Atlantic to start their new life together. Actually, she wrote another page, but ripped it out of the journal, as if those thoughts were not to be recorded. When the couple returned to America, there was still precious little to Colorado Springs, and Glen Eyrie was not even started. So it may have been solely for practical reasons that Queen declared that she'd rather remain with her father in Flushing, leaving Palmer to continue west without his wife.

Queen was still in Flushing a full year later, when Palmer decided he could wait for her no longer. He arranged for Queen to take the Kansas Pacific to Denver, and then ride on to Colorado Springs on Palmer's freshly completed Rio Grande tracks. When she arrived, however, Palmer was not there to greet her. He was off attending to his railroad. In her distress, she could hardly bear to look around. Rose Kingsley had accompanied her there, and in an account of her travels, she described what awaited them:

*You may imagine Colorado Springs as I did, to be a sequestered val-
ley, with bubbling fountains, green grass, and shady trees, but not
a bit of it. Picture to yourself a level elevated plateau of greenish-
brown, without a single tree or plant larger than a Spanish bayonet
two feet high, sloping down a quarter mile to the railroad track and
Monument Creek, and you have a pretty good idea of the town site
as it appears in November, 1871.*

While the town itself may not have been fully built, it was clear
that preparations had begun. Future streets were marked off with long
furrows turned by a plough, and a few houses were going up here and
there. Glen Eyrie, however, had scarcely progressed beyond a foundation. Once the General returned, he and Queen lived together "in a
picnic way," as Kingsley put it, in a loft above a stables across from the
unfinished castle. When then invited Kingsley for tea, she noticed the
cups had no saucers. That winter, temperatures rarely rose above zero,
but it was clear that Queen was trying very hard to make the best of it.
She committed "most perseveringly" to teaching at the local school,
and she joined with Kingsley to start a choral society. Queen herself
sang some ballads in its first performance to "rapturous applause."

Glen Eyrie in 1872.

Unfortunately, when Glen Eyrie was finally done, it proved to be less than the towering stone castle she'd expected. Trying to reserve funds for his railroad, Palmer had limited the structure to a wooden, Tudor-style mansion of fourteen rooms with plans to expand it later. Not a shack, certainly, but not a palace, either. It got crowded when Queen invited her father, Ellen, and their brood to live on the second floor, and Palmer added his mother, the prim Quaker, to the mix. Queen also invited her grandmother, the one who'd raised her, but she fell ill on the train ride from Cincinnati and died a few weeks after her arrival. She was buried beside the castle, the grave marked with a sandstone rock.

By early 1872, Queenie was pregnant with her first child. Not wishing to deliver a baby on the frontier, she asked Palmer to take her back to New York, a request he honored, although he immediately returned to Colorado Springs, so it was there that he learned by telegram that he had a daughter, Elsie. Instead of rushing back East to join his wife and child, he left for Mexico to pursue his railroad interests.

When Queenie finally brought Elsie to Colorado Springs in June the following year, life did not improve. To give her some space, Palmer moved her father and his family to the new resort town he'd created, Manitou, a few miles away. William Mellen died there that fall, plunging Queenie into a fit of mourning; she'd always been devoted to her father. After the panic of 1873 hit, finances grew so tight the General had to rent out Glen Eyrie and move his little family into a modest temporary house in town. Though their money situation improved as the local economy stabilized, things did not get much better for the Palmer marriage.

The General did dote on little Elsie, though. Skinny with bangs drooping over her doe eyes, Elsie was a tomboy, toddling about in chaps and a cowboy hat, squealing with delight when her father swept her up onto his horse to go cantering off. Busy as he was, Palmer always made time for his Elsie. Together, the two of them schemed of ways for Papa to play hooky. "Our mood was a very happy one," Elsie recalled later. "We made up rhymes about the stealing of these hours that should have been spent in town." Palmer let Elsie stay up late into the night with him, reading together and chatting. "Never, not even in the busiest, a slave of time, in moments of relaxation he had—in an extraordinary degree—the gift of enjoying the passing hour, and of making others enjoy it, too."

All the while, Queen was slowly distancing herself from her husband and his treasured Colorado Springs, a situation that must have been agonizing for Palmer. He'd moved heaven and earth to turn the town into the place of his wife's dreams, but these were never her dreams at all. The whole business was an affront. She sensed that the town had been dubbed "Little London" out of some light mockery—the place was never particularly English, more a kind of uneasy Anglo-cowboy hybrid. It had too many cricketers who'd paid to be "presented" at the royal court to gain a patina of nobility, and too many pretentious expats like the insufferable Francis Cholmondeley Thornton, who served as general manager of the new hotel that Willie Bell had built in Manitou. Not to mention the hideous rattlesnakes, so many that Queen's twelve-year old cousin shot five of them in a single year.

A gossipy New York sophisticate, Frances Wolcott, observed that people came to Colorado only after a "loss of health, wealth or reputation." In this, she was thinking of an "intellectual, great, but humorless lady from Fifth Avenue" who came for an extramarital fling with an "aesthetic Englishman" and then left him with a collie to remember her by. Wolcott spotted Queen one Christmas Eve at table beside a tedious English gent "molded in Bond Street's best evening attire," who was boring her to tears with some story of "a St. Bernard dog and an express wagon." Queen was listening, Wolcott noted, with a frozen smile.

Still, she threw herself into her cultural interests, taught school, played the loyal stepmother/cousin to all the little Mellens, bestowing on them pets and ponies, and hiking about. "Queen Palmer climbing stock in hand, rebellious curly hair flying, cheeks aglow, moving as on winged feet, was the spirit incarnate of inaccessible heights," Wolcott recalled, admiringly. "Hats she scorned. She laid claim to beauty as her right of heritage." It was all quite valiant, but it didn't seem that Palmer noticed, he was so preoccupied with charting his railroad's route south. Even when he was home, he was always busy with something that didn't involve her. Before long, Queen took over the entire third floor of Glen Eyrie to make their bedroom hers exclusively, a private space for books and music where, Wolcott says, she "never permitted any but an invited friend to enter." It's not clear if her husband made the list of invitees. Elsie remained an only child.

Wolcott's account suggested that eventually Queen's attention may have strayed to another, a Count Otto de Pourtalès, the rakish son of a Harvard scientist whom Wolcott dubbed the "Mute Seraph." She praised him as a horseman and dancer, unable to resist adding that he was a "born lover" who wooed by singing songs of unrequited passion "as persistently melancholy as the call of a mourning dove." To Queen, the Count's soulful music must have been very pleasant to hear in her private chamber.

Once, when the Palmers were riding by carriage through the mountains to pay a call on the Bells, Queen finally let her anger show. A sudden gust of wind had blown red dust all over them and Queen demanded that Palmer stop the carriage. She jumped out to thump all the horrible dirt off her lovely pleated dress and threw herself under the tossing branches of a mahogany bush to thrash about, screaming. Palmer must have been used to such outbursts, because he made no move to console her. When she returned to him, she was composed, disturbingly so. "I was in a furious passion as if the wind were a person," she curtly informed him. "So I lay kicking and screaming as if I were crazy." Palmer may have found the explanation even more alarming than her behavior.

In 1875, the General brought Queen with him to Paris in a failed attempt to raise capital for his railroad. For the trip, she tried to keep a diary as she had before but ran into trouble with the first sentence. Unable to find the right pronoun for herself, she settled awkwardly on the royal "we." "Four years ago, we crossed the ocean when we began a journal as so many people do—thinking that we would keep it up as long as we travelled!" But that "we" was evidently a torment, since it implied a reference to her and Palmer both. And things got even more confusing from there. "We have been travelling nearly constantly since, and the journal so bravely begun came long ago to an untimely

and abrupt end!" Have "we" been traveling? Or has only *he* been traveling, generally without her? And the "untimely and abrupt end" to the
previous journals made it sound like it somehow ended on its own, not
that she ended it.

But the muddle touched on what was evidently a core dilemma—
was she part of a couple or not? Ultimately, she abandoned that "we"
for a rather lonesome "I" as she notes that, in starting this fresh journal
"I begin again alone." But, unable to face the prospect, she decides to
"close the book in disgust." Thus ends the journal.

When the visit to Paris was over, Queen remained on the East Coast
with Elsie for the next three years. Palmer was all alone in Glen Eyrie,
spending most of his time in his study, which he called "The Grotto."
Dark and cavernous, the room had severed tree limbs arching up to the
ceiling, and a massive stone fireplace that seems to have been extracted
from a rock face to fill the entire wall. In a photograph from the period,
he stares dazedly into the fire as if in an opium haze, the sheets of railroad plans lying all about him, with more designs propped up behind,
as though he just can't bear to think about them.

General Palmer in "The Grotto."

CHAPTER 9

The Battle Is Joined

WHEN STRONG JOINED THE SANTA FE AS GENERAL MAN-
ager in November 1877, he'd immediately made clear to Nickerson
that, while he understood the appeal of some silver mines past Pueblo,
he thought the benefits would be too short-lived—and the move might
leave the railroad stranded in the mountains, with no good way west
from there. Strong was more inclined to follow Palmer's original thinking that the way west ran south over the Raton Pass.

He hired a surveyor, Ray Morley, to find the best train route over
the pass and into New Mexico. Just twenty-six, Morley was a jack-
of-all-trades with a magnificent pair of mutton-chop sideburns. The
orphaned son of a destitute Massachusetts farmer, he'd made a career
out of scrappy improvisation, making him perfectly suited for Strong's
task. He had signed up to join the Union army at seventeen, a year too
young, but to avoid a lie he had written the number eighteen in chalk
on the sole of his boot. That way, when he told the recruiter he was
"over eighteen" it was technically true. He started in railroading under
General William Tecumseh Sherman, who put him to work restoring
various battered Union railroad lines to transport their soldiers through
the South. He was also charged with sabotaging the Confederate lines
by lighting red-hot fires along the tracks to melt the rails into "Sherman's hairpins" that could be wound around nearby trees.*

---

* The effort revealed the fact that the Civil War was, in large part, a railroad war. The

Ray Morley.

After the war, Morley fell in first with General Palmer when he did some surveying for the Kansas Pacific. Impressed with the young man, Palmer hired him to oversee the tunneling into some silver mines on the Maxwell Land Grant and Railway Company after Palmer became its president. When Morley handled the work successfully, Palmer had him take charge of the squatters' land claims, too. Eventually, he had Morley manage the entire estate.

By then, Morley had married a highborn Iowan, Ada McPherson, whose determination to curb the drunken rowdiness of the area led to her founding a local chapter of the Women's Christian Temperance Union. The two lived in some grandeur in the Maxwell mansion,

---

railroads were essential for supply lines and the speedy transport of troops. In rushing vast numbers of soldiers into battle, a railroad could give its side the edge, but also drastically increase the carnage. While the South might have had the better generals, starting with Robert E. Lee, the North had the better railroads. They were longer and more centrally coordinated than the short, scattered southern lines. And in the end, this proved decisive.

where Ada delighted in the roomful of stuffed, exotic birds supplied by the London financiers. But after Palmer left to concentrate on his new railroad, Morley found the soggy gents of the English syndicate increasingly annoying. He agreed, as a sideline project, to help Palmer push his line into the La Veta coalfields, assisting with the surveying on McMurtrie's remarkable Mule Shoe Curve, but when the work was done, he happily left the clubby Rio Grande and returned his full attention to the Maxwell Estate and Ada.

That's when Strong came along. He quickly saw the possibilities in using Morley to survey the pass, but knew the matter had to be handled with extreme sensitivity, lest word get back to Palmer. He put out feelers, careful not to say too much too soon. Would Morley be interested in doing a little freelance work? A bit of surveying, actually? In the Raton Pass? In secret?

Morley assured Strong he'd be willing indeed, and he knew just how to do it with the necessary stealth. He'd go about his surveying work disguised as a Mexican shepherd, one of several in the valleys up there, wrapping himself in a Spanish-style serape, with a black slouch hat pulled low over his forehead, to follow his sheep about the heights while he slyly made his surveying notes. No one would ever suspect.

After his years with the Maxwell Land Grant, Morley already knew the territory—but more important, he knew Uncle Dick Wootton, the crotchety owner of the pass. When he revealed his plan to Uncle Dick, emphasizing the need for discretion, the codger was surprisingly accommodating. He must have realized a train would be the end of his toll road, and likely of his hostelry, too. But he couldn't hold off the future forever. Nearing sixty, Uncle Dick was tiring of life in the mountains with so many bandits and desperadoes coming through. Plus, he'd spent enough time in Trinidad saloons to share his drinking friends' fury at the Rio Grande for trying to kill off their town.

He was fine with having Morley dress up as a shepherd to wander about his pass to find a train route. He'd even supply the sheep. If the company decided to build, he told Morley to send someone to see him about the rights.

Morley roamed about the pass with his sheep well into December, braving the snow and the cold to determine all the inclines and out-croppings for the best train route through. He set up camp there each night, furtively recording his impressions in a sketchbook by the fire.

There was a problem, though. However Morley measured it, the top of the Raton Pass was simply too steep for a train to cross. He couldn't see any switchbacks that would stay below the four percent maximum grade. To get across, the Santa Fe would need to put in a tunnel just down from the ridgeline, but even that would be tricky.

He delivered this news to Strong's hardy engineer, A. A. Robinson, who passed it back to Strong. Strong, in turn, relayed it to Nickerson in Boston to run by his board. The word came back along the same channel. The Santa Fe board would not authorize Strong to pay for any tunnel into New Mexico.

Strong, however, was not about to relent. He told Robinson and Morley to find another way. Together, they did. They'd use a "shoo-fly."

A shoo-fly is a type of switchback, but with a pronounced zigzag that allows a train to ascend a steep mountainside. Sketched out, the route looks like a child's drawing of the edge of a Christmas tree; each curved-up branch is a shoo-fly. To ascend, a train fires up its engines to run as high as it can on the bottom "branch." When the going gets too steep, the engineer throws the train into reverse to run back up onto another set of tracks that are set at a higher elevation than the previous one. After rolling back as far as it can, the train then charges up the mountain once more, this time on another set of tracks that are placed still higher, and so on and on, forward and back and forward again, until the train catapults itself over the top.

There was nothing like it anywhere in North America, but Robinson had heard of its use in Europe. Morley hoped the Santa Fe could get by with two such shoo-fly branches, but when he ran the numbers on the train's weight, the locomotive's thrust, and the track inclines, he realized that one locomotive could not provide enough power. He tried to add a third shoo-fly, but there wasn't room, so he added another locomotive. When that wasn't enough, he threw in a third. It worked.

When the two men sent Strong their plan he didn't hesitate. If they thought it would work, that was enough for him.

On December 17, 1877, Morley's work in the pass was complete. Strong's secret had held. To file his map with the local authorities, he enlisted the Santa Fe's lawyer, a Mr. Gast, to do the necessary paperwork and examine the records of any previous applications to make sure that Palmer had not beaten them to registration. Gast discovered that the General had indeed produced a line of levels at the pass way back in 1872, but for some reason had never completed a survey. If the General was to have a shot at securing the pass, he'd have needed to furnish a full map of his route within five years, and now, at the close of 1877, that time had expired.

Armed with all the necessary maps, Gast gained proper certification for the Santa Fe's claim and prepared to buy the land from Morley, who held it in his capacity as the executive officer of the Maxwell Estate. But none of it could happen until the Santa Fe came to terms with Uncle Dick and got the first shovel into the pass. Once the rights were secured, he'd have Morley build a Santa Fe line down from La Junta, one stop before Pueblo, to reach Trinidad and continue on to Raton from there.

But first Strong needed one last thing: a railroad charter from the territorial legislature, allowing him to bring a railroad into New Mexico. To secure one, he enlisted the politically connected Don Miguel Otero, who had served as a New Mexico legislator.

The legislature met in Santa Fe at the Palace of Governors, a long colonnaded row of dignified rooms that ran along the city's main plaza. The body was made up of two dozen wealthy traders, ranchers, and financiers. Having been one himself, Otero moved very easily among them, making discreet inquiries about a charter for the railroad.

The news Strong received from Otero was not good. Another train had beaten him to a charter, and then pushed through legislation to prevent another train company from getting one. The Southern Pacific, the formidable new California train network that had grown out

THE BATTLE IS JOINED 113

of the Central Pacific half of the Pacific Railroad, had been building east from San Francisco with the idea of eventually creating a new transcontinental all of its own along a southern route through Arizona. Just two months before, the SP had broken into New Mexico, immediately pushing through legislation that imposed impossibly heavy capital requirements on any line that wanted to come in after them. This move was less to protect their New Mexico territory than to preserve their monopoly on California. If they kept other trains out of New Mexico, none could get into their state from the southwest.

Determined to find a way in, Strong told Otero to scrutinize the legislation for possible loopholes, and sure enough, his ally found one. It seemed the SP's legislation had failed to repeal the previous version of the law, leaving it fully in effect without any restrictive capital requirement. Strong was free to send the Santa Fe into New Mexico after all.

Strong had Otero get his old friends in the legislature to establish a charter granting the Santa Fe permission to enter the state, and while they were at it, add language freeing the Santa Fe from paying any taxes in New Mexico for six years. All of this had to happen on the sly, and fast, before the SP realized its mistake or the Rio Grande learned of Strong's efforts—and for an appropriate fee, of course. When the deed was done and the prized charter was in his hands, Otero slipped swiftly away.

But the Santa Fe couldn't make use of its charter, or execute its rights in the Maxwell Land Grant, until Strong could persuade Nickerson of the wisdom of doing so. This posed a challenge, since Nickerson was determined to go west, not south. It took several rounds of anguished pleading from Strong before Nickerson would even think of taking the matter to the full board.

Fortunately for Strong, the General had not gotten wind of his rival's intentions. He was still inclined to go west into the coalfields, not south into New Mexico, and at that point was thinking of driving through to that Trinchera Estate deep in the mountains past La Veta.

Any day, he'd send McMurtrie to Fort Garland to start plotting that line, with the possibility of reaching Santa Fe from there.

Or so it appeared. In late February, while Strong was still pleading with Nickerson, word came from Morley that some men from the Rio Grande had asked if he'd be interested in doing a survey for them in the Raton Pass. McMurtrie had spoken very highly of Morley, apparently. When Robinson passed this news on to Strong, he assured his boss that Morley gave the Rio Grande emissaries "no satisfaction." To make sure he never would, Robinson put him on the Santa Fe payroll. Still, Strong smelled trouble.

Things got worse a few days later when Robinson learned that McMurtrie had gone ahead without Morley and started sending survey- ing equipment from Fort Garland into Raton by way of El Moro. Had the General changed his mind? After all the Santa Fe's careful, clandes- tine preparation, Robinson was terrified that McMurtrie would beat him to the pass after all. "They will try and get a force in ahead of me," he informed Strong by express rider. "I don't want to be out generaled in this game without at least having the satisfaction that I used every avail- able measure to win." He begged Strong to let him get started digging at Raton right away, before it was too late, and on February 26, Strong cabled Nickerson in Boston to demand immediately authorization.

After a quick canvassing of the board, Nickerson agreed, but al- lowed Strong to spend only $20,000 on the project, with the idea of holding off on any construction until sometime "in the spring."

The spring? Strong needed to hit Raton now. And just $20,000? He'd already spent that much. No matter. He'd get more money later and there was not a moment to lose. He fired off a coded message to Robinson at Pueblo: "SEIZE AND HOLD THE RATON PASS."

Back at Glen Eyrie, Palmer finally cast off his enervating gloom over Queenie, and fixed his mind on the matter at hand. He'd always assumed that if Strong was going to build down to the pass, he'd start out from somewhere along the existing line that ran from Kansas into Pueblo—otherwise, he'd have a terrible time getting his rails and equipment in place. He'd seen no sign of any such construction, or any indication Strong intended to build farther west past Pueblo, either. Yet Strong had been hired to build. So why was he not building? For a long while, it suited the General very well not to think about any of this. But now it grabbed him and would not let go. It was frightening, all this silence. Palmer had picked up no word of any activities to secure a charter in Santa Fe, nor at the local office near Raton where he needed to file his map. And he hadn't any idea about Morley's surveying. All he knew was that Strong had been cabling Boston with increasing frequency. His men had not been able to decipher the messages, but the quickening pace of the exchanges was alarming. Was he hiding something? Was he in fact right now preparing to build somewhere, but quietly, secretly, out of sight—only to spring it on Palmer when the deed was done?

It was the Raton Pass. It had to be. Strong was going for it, the General just knew. He was going to take from him what was rightfully his. The Southwest was his, it had always been his. To claim it, Palmer had to move *now*. If Strong wanted it, he wanted it more than ever, and Palmer was suddenly deathly afraid Strong was about to take it. He didn't think it. He *felt* it. Sometime in the morning of February 26 the General's anxieties got the better of him. He cabled McMurtrie to drop everything at Fort Garland, go to Raton, and start digging. He'd figure out about the rights later. Take possession first. McMurtrie cabled back that he'd go to Pueblo immediately and take the first train south from there. Robinson's train.

⊳⊷⊶⊙⊷⊶⊲

When the two men disembarked at El Moro shortly after ten, both were tired from the tensions of the day and the long, rattling train ride. And the chilly, snow-covered town was lit only by a sliver of moonlight. McMurtrie decided to catch some sleep at El Moro's sole hotel and start for the pass at daybreak, but Robinson went straight to Morley, who was there to greet him at the station. He hustled Robinson off to the buckboard he had waiting, and with a flick of his whip, the two men clattered smartly away.

Morley had already assembled a work crew from Trinidad who seemed eager to stick it to General Palmer for trying to strangle their town. He had the men wait outside of town, their wagons loaded with tools and rails. With a wave, he led them all up Uncle Dick's toll road to the pass. It made for an eerie two-hour climb in the moonlight. A storm had dumped nearly a foot of fresh snow a few days before, but warm days had reduced it to a few inches of slush as they rolled along. In the hush, Morley and Robinson listened for any sounds of McMurtrie or his men chasing after them. But everything was quiet.

It was well after midnight when Robinson and Morley finally reached Uncle Dick's hostelry. Its windows were black in the dead of night and Robinson had to give the door a good pounding before a groggy Uncle Dick came down with a candle. He'd been up late hosting a dance and only just retired to bed.

When he saw Morley with all the workmen, he knew why. He'd expected the surveyor would be back for the rights before too long. Midnight wasn't the best time for a negotiation, but Uncle Dick was done with his hostelry and his toll road and agreed to talk. Strong had authorized Robinson to pay up to $50,000 cash for the rights (so much for Nickerson's $20,000), but the old man asked only for a lifelong stipend of $25 a month for him, his wife, and their daughter. (In gratitude, Strong later threw in a free lifetime pass for all of them to travel wherever they wanted on the Santa Fe railroad.) After Wootton and Robinson shook on the deal, Uncle Dick enlisted some of the dancers who'd bunked in at the hostelry after the party to join him at a spot along the route. There, Robinson handed Uncle Dick a pickax to do the honors. He stuck the first blow into the icy ground to pry out the first clump of dirt that certified the Santa Fe's claim. Robinson made a note of the exact time—two o'clock in the morning—to make clear that the Santa Fe had indeed been first to the pass and then sent men farther on to start in at all the key points of the route, with special emphasis on the entrance and exit of the tunnel that Morley had identified for future use, so they could make an undisputed claim to the full route over the pass and into New Mexico.

Soon twenty-five men were digging by lamplight. And not just common laborers, but, as Robinson told Strong, "several merchants and professional men" from Trinidad. They landed their picks and shovels with special fury, striking blow after blow against the despicable General Palmer, who had treated the citizens of Trinidad in such a "shabby manner." Robinson was only too happy to give the put-upon Trinidadians a train. "The citizen will stand by me," Robinson assured Strong, "and I am sure I can hold my position."

The test came before dawn. After a few hours of fitful sleep, McMurtrie could wait no longer to hit the pass. At that hour, he found it difficult to round up workmen, even though the people of El Moro depended on the Rio Grande for their livelihoods. By the time he was able to get a crew to the pass, McMurtrie was stunned to find Robinson's crew already hard at work. He unleashed a torrent of harsh words at Robinson, adding some more for Wootton once McMurtrie realized he was complicit in the theft of what he insisted was the rightful property of the Rio Grande. Seeing so many men from the Santa Fe busily at work, McMurtrie saw no point in putting in any of his own. Instead, he sent engineers scurrying back toward El Moro to find key points along there to claim. Fortunately, Robinson had his own engineers in place there already. "They found they were beaten all round," a triumphant Robinson told Strong.

McMurtrie himself climbed up to the summit of the pass to look around for some section he could claim but came down ashen-faced. His assistant shouted at Robinson that the Rio Grande would build its tracks over the pass come "hell or high water." Still, McMurtrie knew the Rio Grande forces were not just late, but outnumbered, and Wootton had the only firearm. There was no question which way he would point it, should the need arise.

That night, McMurtrie continued to scout about for another route for the Rio Grande, hoping that a Chicken Creek line might be possible, but Morley had done his work too well. By morning, the Santa

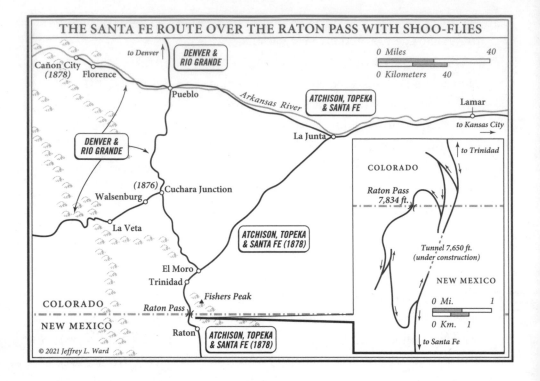

Fe crew had cleared a length of roadbed and then laid down enough trestles to hold the line's first pair of standard gauge sixteen-foot rails. As far as the Santa Fe was concerned, the deed was done.

The General, monitoring the developments from Colorado Springs, refused to accept defeat. For a week, he made McMurtrie stay on in the pass, hunting in increasing desperation for another route. Now that Strong had claimed Raton, Palmer wanted it more than ever.

He kept at it until the middle of April, when some Colorado newspapers reported some stunning news that changed everything. It now appeared that Leadville, one of the mining towns up in the Rockies northwest of Pueblo that had interested Nickerson, had emerged as one of the greatest silver producers in the West. That was exciting for the miners and mine owners, but it would be electrifying for the first railroad that could get there.

Strong and the General both knew the stakes, too. The only route into Leadville ran through the Royal Gorge, whose towering cliff sides, cut by the raging Arkansas River, angled down to an extremely narrow passageway at the water's edge, with room for only a single set of tracks.

The Raton Pass was one thing, the Royal Gorge would be another.

The Royal Gorge

PART TWO

THE ROYAL GORGE

▸◂▸-◦-◂◦▸-◦-◂◦▸-◦-◂◦▸◂

CHAPTER 10

Precious Metals

ORIGINALLY, IT WAS GOLD—OR "COLORS" AS PEOPLE said—that was found up along the Arkansas River west of Pueblo in the fall of 1859. The prospecting did not begin in earnest until the heavy snows had finally melted away the following spring, and by then wave upon wave of hopefuls climbed ever higher into the mountains in search of the richest veins of gold. By summer, a full ten thousand people—a third of the population of all of Colorado—had clambered up into the windswept, rock-strewn meadows of a place called California Gulch a full two miles up. One of the gold bugs was Palmer's man, Alexander Hunt, coming in by oxcart from Illinois. Most of the prospectors lived in wagons, their clothes, dinner plates and other belongings stuffed underneath, if they didn't turn them into ersatz hotels for eager newcomers piling in on horseback or on foot. The population influx was enough to carve Colorado out of Kansas as a freestanding territory.

Ultimately the miners pulled out $2.5 million in gold, the equivalent of $100 million today. The rush ended in the 1870s, but Alvinus B. Wood and "Uncle Billy" Stevens stayed on, thinking there might still be valuable traces of gold left in the gravel of the stream beds. To flush them out, they started the Oro Mining, Ditch and Fluming Company, diverting a flood of water from the Arkansas down an eleven-mile sluiceway into the gulch. It cost $50,000, but they soon made their money back—and then made more when they decided to investigate the strange, lumpy, dark ooze that didn't entirely wash out. Curious about this "black mud," they hauled a ton of it over the Mosquito Pass in barrels to a chemist in Alma who discovered that a quarter of it was lead. At six cents a pound, that was good—but the ton also yielded fifteen ounces of silver, which was far better.

While gold has always had an unshakable hold on the human imagination, silver's allure has been less profound and more intermittent. In 1874, when Wood and Stevens founded their silver company, an ounce of silver fetched a dollar while an ounce of gold went for sixteen. Although silver is about twenty times as plentiful as gold, it's much harder to find and infinitely harder to extract. And while gold comes in a nice, solid lump, unmistakable in color, silver is rarely obvious, usually a hidden component of an otherwise useless ore that miners call "muck." Up in the gulch, the ore was lead carbonate. It could take the form of the gravel, which Wood and Stevens washed out, but it was secreted in far greater abundance in thick slabs of rock well underground. Either way, the ore needed to be yanked out and smelted in a blast furnace that melted off the worthless parts and left the gleaming silver behind.

Its price, curiously enough, fluctuated partly by act of Congress. The Constitution included a provision that paper currency could be backed by silver as well as gold, meaning that citizens could trade a paper dollar for a dollar's worth of either metal. By the Coinage Act of 1873, the year of the Panic, silver was summarily dropped not only from the paper currency, but from the silver dollar. Like politics generally, this reflected the class divide, but it also reflected the emerging geographical divide as the country spread west.

The notes were held almost exclusively by eastern moneymen, but the interest was paid in taxes by everyone after the war, which had left the country in immense debt. That created a political revolt. The eastern financiers wanted the public debt to be held in "hard" dollars that were backed by gold alone, since its relative scarcity would drive up their dollars' value. Westerners wanted to water down the currency by adding silver, creating "soft" dollars that would make the debt cheaper for them to repay. Instead of resolving the matter, the Coinage Act turned West against East for the rest of the century.

In 1878, a panicked Congress repealed the act to restore silver to the currency, overriding a veto by President Rutherford Hayes, who sided with the moneyed interests. As money softened, silver prices soared, but when gold partisans roused themselves to dump silver again in 1893, money hardened and silver prices tumbled, this time ruinously for silver producers across the West. In their desperation, silver backers named the otherwise obscure William Jennings Bryant as the Democratic candidate for president in 1896 after he famously implored the US government not to "crucify mankind on a cross of gold." He was opposed by Republican William McKinley, stolid representative of the eastern financiers, who saw the gold standard—meaning a dollar backed by gold alone—as essential to American prosperity. The issue so gripped the nation that eighty percent of the electorate turned out to vote. After McKinley won a narrow victory, he duly banished silver from the currency in 1900, leaving only gold until 1971, when it was dropped from the currency, too. Now the dollar is backed only by itself, only as valuable as people believe it to be.

In 1874, however, the heavy politics was still in the future, and silver was still plenty desirable just for its own sake. Wood and Stevens kept quiet about their discovery as they snapped up dozens of abandoned claims in the gulch and set up a small-scale smelting operation to pull out the silver. The earlier talk of "black mud," however, had attracted the attention of August R. Meyer, a graduate of the world's preeminent institution of metallurgy, the Freiberg Mining Academy in Saxony. He'd been sniffing around the mountains and wanted in. He hired workmen to dig out three hundred tons of ore from a mine of his own and haul it down the mountain by oxen to Colorado Springs. There, he had the ore loaded onto Rio Grande trains to carry to Denver, where it was transferred to Kansas Pacific trains to cross the prairie to the heavy-duty smelters of the St. Louis Smelting and Refining Company in St. Louis. Altogether, it was a trip of almost a thousand miles.

Meyer's ore did indeed yield silver, but the massive transportation costs severely cut into his profits. Undeterred, he sent another fifty tons of a higher grade of ore to St. Louis. This time, the quality so impressed Edwin Harrison, the smelting company president, that he decided to put a smelter of his own in the gulch.

When Wood and Stevens heard about this, they figured they'd better scale up and bring in a lot more workmen. A silver mine is not just a hole in the ground. Instead, it resembles an underground mansion of several floors, each one spreading out through a number of damp, dirty, low-ceilinged rooms to allow miners to get at the silver that runs out sideways at different levels. It's a huge job to dig it all out with pickaxes. Since virtually all the gold miners had left the gulch, Wood and Stevens had to put the call out to workmen from faraway Detroit and bring them in by a special Rio Grande train to Colorado Springs.

They arrived in April 1877, to a white-out blizzard. No wagons could carry them from Colorado Springs up into the gulch, and so the Detroiters had to climb for miles on foot, carrying all their belongings through snow that started out at their waists. By the time they reached the Oro Company offices, it reached their armpits. Nonetheless Stevens and Wood sent them onward for another mile to various stables, blacksmith shops, and sawmills, where they were expected to bed down on the floor.

This did not make for a cheerful labor force, and things got no better when the work began. Wood and Stevens set the men's wages so low and their boarding costs so high that they barely broke even. After a few weeks, they all wanted out. When someone spotted Stevens and Wood sauntering down one of the slushy streets of the gulch, a mob of angry miners gathered to chase them. In a panic, the two owners ducked into the gulch's one general store, bolted the door behind them, and hid behind the safe in a back room, hoping the threat would pass. It didn't. The outraged men banged on the doors and windows, fired their guns into the air, and displayed such fury that the store's teenage clerk figured he'd better unbolt the door to let them in before they destroyed the place. The men burst inside, firing the guns at the ceiling, and demanded that Wood and Stevens show themselves. When they finally did, the miners unspooled a length of rope and gave their two employers a choice. They could pay them handsomely to return to Detroit or hang from the nearest tree. Stevens and Wood paid up. The men all left for Detroit.

Unnerved by the incident, Alvinus Wood sold out his half of the company to Levi Z. Leiter, a former partner of the retailer Marshall Field, for an impressive $40,000. When word of such a bell-ringer sum got out, it proved such a come-on that others clamored to get in on the action, not the least of them the Colorado territorial governor, John Routt. Largely abandoning his gubernatorial responsibilities, he snapped up a mine of his own, the Morning Star, and worked it himself.

Stevens and Leiter, meanwhile, continued to expand their holdings until they hit trouble in the form of a disgruntled miner named Williams. Described by the *Chronicle* as a "brave man, perhaps a desperado," this Williams rashly sank a two-hundred-foot mining shaft right down through one of Stevens and Leiter's many claims, and then barricaded the opening to keep possession. In response, Leiter and Stevens hired twenty well-armed "fighting men" to reclaim their property. When Williams responded by stuffing the mine with his own gun-toting ruffians, Stevens and Leiter lit torches to toss into the shaft. At that, Williams cleared his men out, and he quit the field, cursing.

Plenty of other prospectors rushed in to replace Williams by more lawful means, but that encounter set the tone for California Gulch for some years to come. For every claim, there was a counter claim, and then another counter claim to that one, fury rising with each round. It seemed that the lust for possession was total. No good thing could be shared. Before long, it took four hundred lawyers to handle all the disputes in Leadville.

>+•>-0-<•+<

Palmer had been following the developments in California Gulch for almost a year before he finally made his move in 1877. He knew Edwin Harrison, the smelter, from his days in St. Louis with the Kansas Pacific, and started pumping him that summer for information about how many loads of ore a train might haul out. The answer was staggering. Harrison was talking fifteen thousand *tons* of ore coming out of the mines every year. That made for a lot of trains. Palmer hurried to the gulch that September with James McMurtrie to look things over. Upon arrival, they spotted not just Colorado's governor Routt but one of its senators, too. Both of them, Palmer noted, were "out with picks searching for ore." A good sign, surely. They also spoke to Uncle Billy Stevens and Leiter, who were even more encouraging. Their Oro Fluming Company was looking to take out a thousand tons of ore a *day*, and their rival Augustus Meyer said his mines were daily producing another five hundred.

Better still, there was another mine at Fairplay, southeast of the gulch, to be serviced by a train. The lush cropland of the region could generate even more business. In a follow-up report, the General excitedly summarized the potential freight haul: "Up—Hay, coal, coke, merchandise, machinery, forage, flour, provisions, groceries, etc. Down—Ore, base bullion—lumber, cattle and beef both ways . . . ."

Looking around, Palmer thought that the gulch might even become another Colorado Springs, but with more spectacular views since it was higher up. "With a railroad, this could be the most attractive summering spot in Colorado," he imagined, and fishermen would be drawn to the Arkansas River since it "could not be exhausted of fish." The dry air would please the consumptives suffering from tuberculosis—if the town was placed well clear of the mines, of course.

These plans were not a sudden revelation. The General had in fact been thinking about a passage into these very mountains ever since his surveying days with the Kansas Pacific. Just as he had at Raton, he'd started a plat in 1872 before turning his attention elsewhere. Now, with McMurtrie, he plotted a more precise train route up to the gulch. It would start at Cañon City, currently the western terminus of the line from Pueblo, and from there climb up through the narrow canyon of the Arkansas Valley, the Royal Gorge. It would make a spectacular ride for the sightseers who were starting to come in, running alongside the

Arkansas River, which tumbled down between two towering rock faces that rose up almost two thousand feet on either side.

For now, the General pictured a train just up to the gulch but knew it could someday run down the far side of the Continental Divide and continue on to Utah to join the Central Pacific at Salt Lake City and create a second transcontinental route, this one passing through growing population centers to generate profitable traffic.

What Palmer had not realized, however, was that the citizens of Cañon City were not at all keen on any venture he would back. Like the Trinidadians, they harbored bitter memories of how the General had so cruelly ended a train line just a few miles short of town. Clearly seeing that California Gulch would amply reward any train that reached it, they resolved to own that train themselves. They quietly established the Cañon City and San Juan Railway, hired the Santa Fe railroad's surveyor to select the route, and entered into secret negotiations with the company to buy them out when the time was ripe. They well remembered that the Santa Fe had come into the original Pueblo, not Palmer's South Pueblo. That made the Santa Fe their kind of train.

CHAPTER 11

"A Game of Bluff"

IT WASN'T UNTIL MARCH OF 1878, A WEEK OR TWO AFTER he'd come in second to Raton, that Palmer realized that Strong might try to seize the gulch's silver riches. The General should have guessed sooner. The latest silver bill from Congress, the Bland-Allison Act, had been passed that year on February 28, restoring silver as a backing of the dollar and unleashing the US Mint to buy silver in quantity. Prices duly jumped and made the news from Leadville all the more exciting, but, caught up with the twin frustrations of Raton and Queenie, the General didn't get moving until his new general manager, Colonel D. C. Dodge, reported a shocking development. Palmer's old friend from St. Louis, the smelter Erwin Harrison, had betrayed him, offering Strong a contract to take out twenty-five tons of ore a day if he could deliver a train to Leadville. That haul wasn't much, but it was more than enough to support a train. In fact, Harrison was already headed to Boston to discuss the deal with the Santa Fe board. "They are determined to get his shipments of ore if possible," Dodge informed Palmer. "Mr. Strong is getting all the information he can with regard to that section and, I believe, intends to make a move in that direction."

Palmer simply could not believe it. Strong *again*? The man had taken the Raton Pass just weeks before. Now he was going to spring for Leadville? No one could move that fast. Or was it just that Palmer couldn't? The Denver papers were all of the opinion that Strong could never get over the Raton Pass *and* build into Leadville at the same time.

One added the suspicion that Strong was engaged in "a game of bluff" at Raton, employing cheap Mexican labor to engage in a charade of building over the pass to lull the General into thinking Strong was so committed there he'd never be able to run a railroad up into Leadville.

The truth was, Strong was not in nearly as good a position as the General to hit the Royal Gorge—but to him, that simply did not matter. The Santa Fe ran only to Pueblo, forty miles from Cañon City, and would require him to build west from there. Yet General Palmer's Rio Grande was already in Cañon City, allowing him to deliver men and materials to the Gorge on his own train.

McMurtrie wasn't so sure that Strong wouldn't find a way. His tracks up into the pass looked all too genuine—and showed that, if Strong had done it once, he could do it again. For his part, McMurtrie was ready to give up. He'd looked and looked, and there was no other good way over Raton other than the one Strong had seized. If it was up to McMurtrie, he'd get into New Mexico via Trinchera just as the General had initially thought of doing. Otherwise, the Rio Grande would just be handing all of the Southwest to the Santa Fe.

Unwilling to commit to the gorge any sooner than necessary, the General decided on a bluff of his own. He'd make a big show of taking the pass after all, incorporating a new railroad, grandly titled the Colorado and New Mexico Railway Company, to cross the Raton at another unstated route. The announcement alone would force Strong to focus on the pass, redoubling his efforts and his expenses, not just to build over Raton, but to build to it from La Junta and past it into New Mexico. With all that going on, surely there was no way he would even think of running miles of track up into the Royal Gorge. Strong would complete his present business before committing to doing more, right? That was only natural. That's what the General would do, so that's what he figured Strong would do. Weren't they fundamentally of the same mind?

A month later, the General was still formulating his plan when he got more alarming news. McMurtrie had learned from spies that a suspiciously large number of Santa Fe men were prowling around the gorge, no doubt preparing to lay track. He hadn't dared investigate for fear he'd be spotted. "I will try to avoid going if possible," he told Palmer. "All my movements are watched, and should I go I am afraid Atchison"—as he called the Santa Fe—"will know of it." That would only provoke Strong to spring into action as he had at Raton. "[He'd] take it that we mean to move in that direction, and to stop us, jump into the Canon and commence work at once."

To McMurtrie it was all too clear: Strong was not mimicking the General. He didn't give a fig what the General had done, and how or why for that matter, only what the General was going to do, so that he might do it first. Once the Santa Fe crews started in, "they could trouble us," he warned the General. That was putting it mildly.

Strong had clearly thought the matter through. He knew that if the General was to get going in the gorge, he'd use the workmen currently at Raton. They'd be the giveaway. Strong told his man at Raton, A. A. Robinson, to keep an eye out for any signs of suspicious activity.

Sure enough, early on the morning of April 19, Robinson noticed fewer Rio Grande workmen in Raton than there had been the day before. He sent some spies to snoop around, and they reported spotting at least a hundred Rio Grande workmen plus horses, mules, and equipment, at the El Moro station, waiting to board a train. When the spies struck up conversations to find out where they were headed, the men all said they were headed to Alamosa. That was past Fort Garland in the direction of Trinchera, suggesting they were building into New Mexico from there. When Robinson cabled the news to Strong at his office in Topeka, Strong did not believe a word of it. The Rio Grande workmen were headed to the gorge. He just knew it. Robinson had to get there first.

"SEE TO IT WE DO NOT 'GET LEFT' IN OCCUPYING THE GRAND CANYON," Strong cabled back, using his term for the Royal Gorge. Go straightaway, he told Robinson, grab some workmen at Cañon City, get to the gorge, and start grading. In such a narrow chasm, the Santa Fe would leave no room for another line.

But when Robinson reached the El Moro station to board a train to the gorge, the Rio Grande officials intercepted him and refused to let him board any Rio Grande train, regardless of destination. He would stay exactly where he was. They cabled Palmer to let him know they had him. The General, it seemed, had the whip hand now—and he would have retained it if not for Ray Morley.

Just then, the mutton-chopped surveyor was aboard a Santa Fe train headed to La Junta, where he was overseeing the construction of the Santa Fe's new line down to Trinidad. Trapped at the station, Robinson dashed off a cable to him with instructions to rush "at once" to Cañon City, collect a crew, and hit the gorge. Unfortunately, Morley didn't get it until his train pulled into Rocky Ford, one stop before La Junta, and couldn't piece out the code until the train had gone several more miles down the track. When he deciphered the message, he stopped the train with a shout, jumped off, and raced back to the Rocky Ford telegraph office to reply. Robinson replied with the breaking details: McMurtrie had been at the El Moro station, loading his workmen and supplies onto a train. Robinson assumed it would take him a while to get them all boarded, and he'd have to transfer them all to another train for Cañon City. Still, Morley must hurry.

To speed to Pueblo, Morley had to get a private train from its station, but the telegraph office was thick with Rio Grande men, so Morley didn't dare put his request through until they broke for lunch at noon.

Darkness was falling when Morley reached Pueblo. He took a hackney to the Rio Grande station in the company's new South Pueblo where, watching from the shadows, he saw that the Rio Grande train had just pulled in. The stiff, soldierly McMurtrie marched about the platform as he angrily directed the unloading of all the animals and men.

Now, how to be first to the gorge? Morley certainly couldn't take the same train as McMurtrie—but there was another way. In Pueblo,

he'd stabled a swift black gelding named King William after the Con-
queror. A marvelous animal, he'd belonged to one of the idiot Londoners of the Maxwell Estate who didn't realize what a beauty he had. Now
Morley saddled him up and rode him through the dark night, tread-
ing carefully along the rough, meandering, unmarked roads. Dawn was
breaking when they finally reached Labran, a few miles shy of Cañon
City, and raced on from there in the daylight at a full gallop.

When he checked the railway station, Morley saw that no Rio
Grande train had pulled in. No McMurtrie, no men. Relieved, he went
cantering through town, calling out for workers to help him build into
the gorge for the Santa Fe. Dozens of men came pouring out of their
houses, only too happy to help. The *Chieftain* reported it took Morley
"but the work of a few moments" to persuade "every available man and
boy in the city to shoulder a shovel, gun or pick" against Palmer. Morley hurried everyone to the gorge, where they immediately laid rails to
claim the route to Leadville for the Santa Fe.

McMurtrie's train was stopped at Labran when someone dashed
aboard with the terrible news about the Santa Fe beating the Rio
Grande to the gorge, sending the Scottish engineer into a towering
rage. When his train reached Cañon City, he sent the company surveyors ahead "on a dead run," the *Chieftain* reported, to start "chaining
and staking the ground" to claim the gorge for the Rio Grande instead.
McMurtrie's men were just a half-hour behind Morley's, but they were
too late. When they arrived it was all too clear: the Royal Gorge had
been taken by the Santa Fe.

For McMurtrie, it was all too infuriating. *Again?* Snapping obsceni-
ties at Morley, he set his own workmen and their mules stomping all
over the Santa Fe's tracks, knocking aside rails and kicking away trestles. Then he put his men to laying the Rio Grande's own rails farther beyond, as if to claim that the Santa Fe was not ahead of the Rio
Grande at all. It was *behind.* So there!

Friendly to the Santa Fe, the *Chieftain* could not contain its delight
at the outcome. Catching Weasels Asleep, or How Morley
Outflanked McMurtrie, ran the headline over its celebratory

account of Morley's midnight ride. The paper compared it to Union general Philip Sheridan's famous Civil War gallop to take charge of the troops at Cedar Creek in the summer of 1864, turning a potential rout by the Confederates into a Union victory that propelled Lincoln to reelection.

While the stakes of this contest were hardly so high, the paper was right about one thing. The battle for the Royal Gorge was on.

CHAPTER 12

Haw

TWO DAYS LATER, ON APRIL 21, IT BECAME ALL THE clearer what the two lines were fighting for. There was another silver strike in the California Gulch, and this one was the biggest ever.

Once a rowdy mining camp of soot-blackened log houses and raucous saloons, California Gulch had grown into a proper town with a drug store and a hotel. At first, people took to calling it Slabtown, but as the silver yields increased they thought the name needed more gravitas. The learned August Meyer recommended calling it Agassiz after the Swiss naturalist. But the popular shopkeeper, Horace A. W. Tabor, pushed the name Leadville, and when the town was incorporated in January 1878, Leadville it was. Tabor became its first mayor.

If ever there was a mayor who fully embodied his town, it was Tabor, his first name usually rendered as "Haw" for his initials. In the great drama of the unfolding West, with powerful interests colliding to transform the landscape, Tabor introduced a rare comic element. Stoop shouldered, ambling, awkward, he gave off a bumbling impression, especially with his mustache, a bushy inverted V that was destined for caricature. But, like Leadville itself, Tabor's lumpen qualities faded from view when he got rich, which was amusing, too. It was as if everyone was blinded by his diamond cufflinks and matching tie pin. The truth was, Haw Tabor was about the luckiest man on earth, the beneficiary of not just one stunning and utterly fortuitous silver strike, but three, each one more stupendous than the last. He represented the

Horace A. W. Tabor.

great hope of the West, the glittering possibility. When Horace Greeley famously advised young men to "Go West," he might have added, "and be made anew." Formerly, the East had been the place of renewal for European immigrants. Now the West was the place for their descendants to achieve a better life, turning a mere shopkeeper into the greatest of the Silver Kings. Haw Tabor became Haw Tabor of the Tabor Opera House, of the Tabor Military Brigade, and ultimately of the United States Senate. To him, though, his greatest accomplishment of all was becoming Haw Tabor of Mrs. Haw Tabor. Formerly Elizabeth McCourt of Oshkosh, Wisconsin, she was, in his eyes, the bubbliest and most delectable young lady in all of Colorado.

Born in Holland, Vermont, Tabor had started out a stonecutter in Massachusetts before moving to Augusta, Maine, where he married his boss's daughter, named Augusta for the town, which she eerily resembled, combining prim earnestness with a superior manner that other men must not have found winning, or she would never have settled for Tabor. At that point, he was little more than a drinker who did some brawling on the side, and she quickly set about reforming him. To try to prove himself to his bride, Tabor took her to Kansas in 1857, but life was no better on the midwestern prairie. Three years later, he moved them to Colorado to get in on the gold bonanza, but missed out on that, too. After Augusta delivered him a baby, it fell to her to support them all. She did some baking, ran a boardinghouse, and opened a general store to profit off the steady supply of starry-eyed but inept prospectors like her husband. Eventually, even Tabor could see he was a flop, and he settled into a quiet life running his wife's store. Genial and unassuming, he got on well with the customers. Affability was probably his one true talent.

His first burst of good fortune came the day after the Santa Fe and Rio Grande first faced off at the gorge, when two unpromising prospectors came into the Tabor store for supplies but lacked money to pay for them. All they owned was a pick, a shovel, and a dog. "The worst played-out man I ever met," Tabor later called one of them, August Rische, a down-on-his-luck cobbler from Missouri. And by then that was saying something. Augusta would have shooed them out the door, but Tabor took pity on these poor sods when they asked if he'd take the $64.75 payment for their groceries in a one-third share of any silver they found. Against all reason, Tabor said sure. He even threw in a jug of whiskey. Departing Tabor's store, Rische and his shiftless partner, George Hook, celebrated with the whiskey, and stumbled on into the silver fields until they both dropped. When they roused themselves, they decided that particular spot, thus far unclaimed, was as good as any to stick in their pick. (So goes the story anyway.) They kept at it for a week, descending a remarkable twenty-five feet down when one of them clicked on something.

They'd struck a thick layer of carbonate that wasn't just unusually but *spectacularly* rich in silver. When the two men knocked out a ton of ore, they smelted out eight hundred dollars' worth, and that was just the start of it. They must have been guided by divinity. As an analysis by the United States Geological Survey later determined, if they'd started digging just a few feet away in any direction, they'd have missed it. They lovingly called their mine Little Pittsburg since Hook had once worked in the Pittsburgh steel mills.

True to their word, the men duly cut in Tabor for his third of the find and then the trio went at the mine in earnest. Soon enough, the Little Pittsburg became not just the biggest producer in Leadville, but in all of the entire United States, kicking out $10,000 in silver a *day*.

That was just the beginning. The other two men sold out their ownership, but Tabor held on to his share and used the proceeds to start the Bank of Leadville with himself as its president. Six months later, he won election as the state's lieutenant governor. Within the year, Tabor accepted a $1 million offer for his share of Little Pittsburgh but took a good chunk of his payment in stock in the Little Pittsburg Consolidated, a publicly traded holding company that now officially owned the mine. He collected *another* $1 million when the share price of that Little Pittsburg Consolidated shot up from five dollars to thirty. This was the kind of world-class luck he had, and it only continued from there.

Tabor was suckered into buying the rights to a mine on Fryer's Hill that appeared to be silver rich, unaware that the only silver in the mine had been placed there by the seller, whose name alone, "Chicken Bill" Lovell, should have raised an eyebrow. Unfazed, Tabor hired men to dig a little deeper—and sure enough hit a body of silver-bearing ore that was even richer than Tabor's old Little Pittsburg. That was the Chrysolite, and it was soon cranking out $100,000 a month. This time, Tabor incorporated the Chrysolite Mining Company himself. When he sold $10 million in stock, the proceeds were payable to him, and he kept enough back that he made millions more when the shares zoomed from five dollars to forty-five. Soon, he started buying up mines all over the lot.

One of them was the Matchless Mine, which, true to its name, proved to be the biggest producer of them all, delivering, all told, $11 million in silver, or over $280 million in today's money.

In Leadville, Tabor did not rise alone. His story was one of many that dramatically increased the appeal of the town as a train destination. What had once been a town of just a few hundred miners was now home to almost fifteen thousand people, and there was talk of its replacing Denver as the capital of the state. Collectively, its one hundred and forty mines were not just the biggest silver producers in the United States, they also produced more silver than all the other mines in the nation combined. Only Mexico outproduced the town of Leadville.

It became such a destination that according to one magazine article, its streets were "filled to the center with a constantly moving mass of humanity, from every quarter of the globe, and from every walk of life." Not just all day, but all night, too, the town brightened by thousands of coal oil lamps everywhere. "It's day all day in the daytime," ran the ditty, "and no night in Leadville." The town's saloons, dozens of them, were jammed with drunken cash-rich gamblers betting on card games and roulette wheels, while piano players thumped out boisterous tunes and painted ladies targeted the clientele, hoping to entice them back to one of the many brothels in town for a little fun.

Leadville in 1879.

The class divide, rarely seen elsewhere in the West, became evident in Leadville as its fortunes rose. Most of the workmen lived in rude huts if they were lucky, otherwise they rented a bed under a vast tent for eight hours at a stretch. For the better off, the town boasted plenty half-decent hotels, cresting at the justly named Grand Hotel. It also had a half-dozen dance halls and some variety-act theaters. For the princes of the city, Main Street featured a smattering of fine Victorian homes with statuary and flower gardens, of which Haw Tabor's was, naturally, the most ostentatious.

But for all its gloss, the town was only a little less uproarious than Dodge City. BRAWL IN A STATE STREET BEER HALL, ONE DEAD, headlined the *Leadville Chronicle*. DANCEHALL GIRL TRIES TO ARSENIC HER WAY FROM LEADVILLE. STREETS FULL OF PERIL. IS THERE NO LAW IN LEADVILLE?

To maintain order, Mayor Tabor relied on city marshals, but a mob ran the first one off after a few days and a junior officer shot his replacement. Finally, Tabor settled on a burly, no-nonsense Irishman, Martin Duggan, who served out a successful two-year term before leaving town in 1880. Tabor then created the Tabor Light Cavalry, outfitting the troops in military costumes of bright blue coats, red trousers, and brass helmets that seemed to have come from a Gilbert and Sullivan light opera. He appointed himself its general, with a uniform featuring fluffy gold epaulets and a steel scabbard inscribed GENERAL H. A. W. TABOR, C.N.A. (It was anyone's guess what the initials stood for.)

It was typical of the double-edged nature of Leadville—wealth on one side, poverty on the other, and crime running all the way through—that the night Tabor opened his opulent Tabor Opera House, a corpse was dangling from a rope across the street, the handiwork of town vigilantes. Nevertheless, the opera house drew such stars as John Philip Sousa, Sarah Bernhardt, Harry Houdini, and even Oscar Wilde. Knowing of Leadville's interest in silver, Wilde read onstage from the autobiography of the only silversmith he knew, the Renaissance Florentine, Benvenuto Cellini. "I was reproved by my hearers for not having brought him with me," Wilde airily recounted later. "I explained that he had been dead some little time which elicited the inquiry, "Who shot him?" To their surprise, Leadville-ians were charmed by this Victorian dandy who wore a tulip in his lapel. He marveled at the town's distinctive cuisine: "The first course was whiskey, the second course whiskey, the third whiskey." Tabor invited him to see his Matchless Mine, and then lowered him down in a bucket to get to the very bottom of it. There, Tabor handed Wilde a silver drill to open the shaft he'd dubbed "The Oscar." Wilde obliged, and later admitted that he'd hoped that Tabor's associates would offer him shares, "but in their artless untutored fashion they did not." Asked later if the good citizens of Leadville were as "rough and ready" as advertised, Wilde thoughtfully replied, "Ready, but not rough. . . . There is no chance for roughness. The revolver is their book of etiquette. This teaches lessons that are not forgotten."

Tabor's transformation was not complete until he moved on from the austere Augusta to the dazzling Baby Doe, twenty-four years his junior. She'd come to town hitched to Harvey Doe, a handsome dullard from their Oshkosh hometown. Baby Doe's attention had already strayed to a dreamy clothier, Jake Sandelowsky, leading to a rift that culminated in her flinging a hunk of ore at her husband's head. She missed, but Harvey Doe got the message. Granted a divorce, Baby Doe swiftly rejected Sandelowsky, too, and set her sights on the Silver King.

Knowing that Tabor liked to drop in at the Saddle Rock Café with his business partner after catching a performance at his opera house, Baby Doe positioned herself at a nearby table, ordered oysters, and, when Tabor and his partner pulled in, made a point of paying attention only to her appetizer. Before long, a waiter brought her a message, scrawled on the back of a theater program. "Won't you join us at our table?"

When she arrived, Tabor ordered champagne, and, although his partner continued to talk business, Tabor had eyes only for Baby Doe. The partner took the hint and gracefully departed. As Leadville's mayor and most prominent citizen, Tabor had the down-low on most everyone

Baby Doe.

in town, and he had the full dossier on the bosomy woman before him, acting so fluttery. He knew about her divorce, and about her affair with Sandelowsky. Baby Doe admitted she wasn't in love with the clothier, but, well, she did have certain financial obligations to him. The champagne flowed, the night went on, and by the time it was over, Tabor had written Baby Doe a check for $5,000 to free herself of any debt.

"Look on it as a grubstake," he told her. Thus was his newest mine acquired. He was sure that it would deliver him nothing but good fortune, too.

⊢•❖•◦•◀•◉•▶•◦•❖•⊣

CHAPTER 13

The Dead-line

GENERAL PALMER WAS NEVER ONE TO GIVE UP. HE HAD
not immediately surrendered at the Raton Pass, after all, but kept at it
for weeks trying to salvage his fortunes. So he would continue to fight
for the gorge, by God, and this time with even greater determination. It
was bad enough to be beaten once by Strong. He would not be beaten
twice. He dispatched lawyers to the local courts to argue that it didn't
matter who came first to the gorge. He had a prior claim from a plat
he'd filed way back in 1872.

But to the General's frustration, he learned Strong was ahead of
him there, too. He knew all about that claim and rushed in lawyers of
his own to say that any efforts in 1872 were irrelevant because Palmer
had "neglected to . . . file the plat of the projected line with the General Land office, in accordance with the act of Congress dated March 3,
1875." By law, the gorge was his for the taking.

Worse, Palmer's history of high-handedness in Cañon City was reason enough for a local judge to send out a sheriff to tell McMurtrie
to stop building through the gorge. When McMurtrie flat out refused,
adding a few flourishes of Scottish indignation, the sheriff frog-marched
him out of the gorge and into a Cañon City jail. Furious at this turn
of events, his Rio Grande workmen turned on the Santa Fe workers
and, with shoves and a few punches, tried to remove them from their
worksite by force. When the Santa Fe men held their ground, the Rio

Grande men gathered up some of their rivals' tools and pitched them into the river.

McMurtrie posted bond to get out of jail and shot off a cable to the General in Colorado Springs apprising him of the developments. Palmer ordered his men to cut off Strong's communications. If Strong was going to try to run this war from faraway Topeka, Palmer would get his men to clip the telegraph and establish a posse to intercept any Santa Fe messages on US mail stages.

To Strong, that posed no problem. He created his own pony express to carry his communications in and out. Then Strong sent out surveyors to make a show of plotting a Santa Fe train to run from Cañon City to Pueblo, thereby doubling the Rio Grande's tracks along a route that Palmer depended on for essential revenues.

In response, the General had McMurtrie rush men into the gorge to stop the Santa Fe from proceeding with any track building. To recruit workers, McMurtrie offered a sky-high rate of three dollars a day, enough to entice some Santa Fe men to switch sides—until Robinson matched McMurtrie's price to win them back.

Workmen poured into the gorge and set up rival camps on the flatland past the mouth of the gorge. By night, they faced each other down, some of them with rifles at the ready, each side daring the other to fire. By day, they laid rails up into the gorge side by side, since there was still room for two parallel sets at the start. The gorge tightened to allow only one beside the roaring Arkansas River farther on.

As they labored in the bottom of the canyon, its walls rising steeply up on either side, Palmer remembered his long-ago fight with the Apache in Arizona. Back then, of course, the Apache had been the ones atop the canyon, and Palmer down below with his men. Reversing that now, he had McMurtrie's man James DeRemer—another one of the hardy Fifteenth Pennsylvania Volunteers who'd started out with Palmer in the Kansas Pacific—send some workmen up the cliff, high above where the Santa Fe team was working. Up there, he'd go the Apache one better to create a series of forts out of loose stones, a few hundred yards apart. Stoutly walled along the front, facing the gorge, they'd offer protection for his Rio Grande soldiers. He had them loosen boulders, sending them crashing down the cliff side, creating a rock-slide to smash the Santa Fe tracks far below. To avoid any reprisals, they'd take cover in the forts. If the avalanche took out any Santa Fe men, too—all the better.

Up went DeRemer's men, and soon after, down came the boulders. Seeing the rockslide, the Santa Fe men scrambled to safety, but the tracks were clobbered.

Ray Morley was in charge of the Santa Fe work crews, and he wasn't any more inclined to take this assault than the General had been. Two could play at this game. Morley had his men create forts of their own across from the Santa Fe side of the gorge, and from there roll down their own boulders to smash onto the Rio Grande tracks. And he armed his men with rifles to level at their enemies across the gorge.

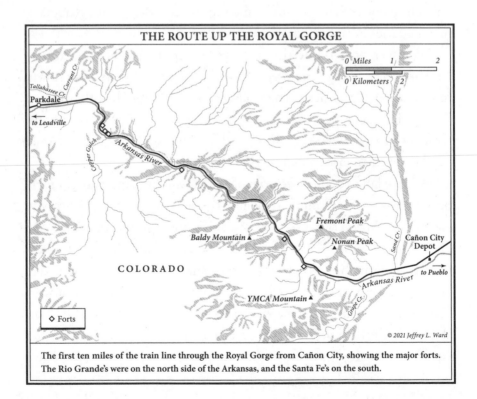

The first ten miles of the train line through the Royal Gorge from Cañon City, showing the major forts. The Rio Grande's were on the north side of the Arkansas, and the Santa Fe's on the south.

Sure enough, once DeRemer saw the Santa Fe guns come out, he did exactly the same. Guns at the ready, each side stared down the other from identical stone forts across the chasm.

It summed up everything about the war between the Santa Fe and the Rio Grande. Each side had goaded the other into acting not just equally badly, but *identically* badly. In their opposition to each other, they'd become indistinguishable from each other. In vying for a single objective, they both became the same.

Determined to come out ahead, Palmer appealed to the federal appeals court, presided over by Judge Moses Hallett, the one who had once sided with him against the Dutch investors trying to send the Rio Grande into bankruptcy. Now that the Santa Fe had found sympathy with the local Cañon City judge, he figured he might have an edge with Hallett. Hallett declared the Royal Gorge matter far too complicated to decide quickly, but for now he ordered both railroads to stop the nonsense. Put away their guns and abandon the forts. He told the Colorado governor to send out the state militia to enforce the order. The troops that had rushed into the gorge drained back to Cañon City.

In a series of rulings that summer, Judge Hallett thought he had the perfect Solomonic solution—split the difference. In his judgment, the Rio Grande had not complied with an act of Congress in 1875 that would have validated Palmer's claim, and so he gave the Santa Fe the right to the best route through the gorge for being the first to build, leaving the Rio Grande to make do with what was left. Where there wasn't room for two separate lines of track, Hallett compelled them to add a third rail to accommodate both gauges.

The General should have been relieved that he was not left out of the spoils of Leadville altogether, but Strong had gotten under the General's skin. He was outraged that the judge was giving his enemy the property and found it unbearable to be forced to take the only route Strong left him. In his anger, though, he overlooked a key fact: Hallett's ruling applied only to the first twenty miles of the run through the gorge up to Leadville. He'd let stand the General's claim to the remaining thirty-seven, declaring that Palmer had properly claimed legal primacy up there in time. For that portion of the route, the roles were reversed, with the General having first dibs, and Strong forced to content himself with what was left. As a result, the two routes would be mirror images of each other. Whatever advantage the Santa Fe possessed up through the first twenty miles was exactly reversed for the remainder.

That was not good enough for the General, and he could not resist trying to gain an edge on his rival. On his portion of the route, above the gorge, he planned to map out a route for his train that would leave the Santa Fe with terrain that would be impossible to build on. "If this plan proceeds," he gleefully informed a subordinate, the Santa Fe "would not [be] likely to enter the Cañon at the Royal Gorge below, because they would not be able to get out above."

To mark the triumph, Palmer had DeRemer nail a railroad tie crosswise to a stake and drive it into the ground at the twentieth mile where the Santa Fe's primacy ended and the Rio Grande's began. It bore the words "Dead-line," a skull-and-crossbones warning to the Santa Fe. To reinforce the point, DeRemer dug two of the most intimidating stone fortresses yet out of the cliffside and stocked them both with Rio Grande riflemen to scare away any Santa Fe workmen who dared venture up.

Finally, the General went on the attack, refusing to allow the Santa Fe to shift any freight onto Rio Grande trains, choking off its cash flow and filing suit to block any new Santa Fe line into Cañon City. That would restrict Strong's ability to import building materials and preserve critical Rio Grande revenues. He then announced grand plans to press on to the city of Santa Fe via the Trinchera route that McMurtrie had favored.

As for Hallett's decision to allow the Santa Fe into the gorge, Palmer made clear he would appeal it to the Supreme Court—and he let Strong know he better not lay another rail into the gorge until that higher court ruled.

Bluff or not, such boldness impressed New York financiers, who rewarded the General with a much-needed cash infusion of $400,000 to start building past the Dead-line up to Leadville.

Strong just watched. By now, the Santa Fe had formally acquired the Cañon City and San Juan Company to build through the gorge. Learning of Palmer's $400,000 to Leadville, Strong created another subsidiary to take the line on up past the twentieth mile and allotted to it a full $6 million. As for DeRemer's forts full of RG soldiers past the twentieth mile, he declared he'd sue the daylights out of the Rio Grande if any of its employees offered so much as an unkind word toward the SF's lawful construction. He flooded that portion of the line with his workmen, hundreds of them. So much for any Dead-line.

And he did not stop there. Strong threatened to shadow Palmer's Rio Grande with Santa Fe rails *everywhere in the state*. All 337 miles of RG track throughout Colorado, not just from Pueblo to Cañon City. He'd run a line up to Denver, into Alamosa, all through the mining country, and even to Palmer's beloved Colorado Springs. He'd plunge a dagger deep into Palmer's heart and twist it round and round and round. Enough was enough.

This time, the General blinked. Much as he wanted to throw his weight around, he had to accept the fact that he didn't have much weight to throw. A baby railroad was in no position to turn vast supplies of capital into endless lines of track, particularly when it was burdened with debt from its existing lines. Both companies were hemorrhaging cash in their fight for Leadville, but in a war of attrition, the Santa Fe had more staying power. Palmer feared that Strong would bleed him dry, but he had one thing still working in his favor: Strong depended on his corporate higher-ups to go along with anything he proposed, whereas the General could do whatever he liked.

Palmer's great hope lay with the Supreme Court, but its decision was many months away. He was absolutely convinced he'd win, but few people agreed with him. Even Willie Bell had his doubts. And if the ruling went against the General, he was doomed.

In October, six months into the war for the gorge, Strong made a stag-gering proposal. He offered to take Palmer's railroad off his hands. Every inch of track, every station and wheelhouse, every employee, every train and piece of equipment. But not to buy it. Strong, frankly, didn't want it that much. To lease it.

To Palmer, it was an offer of breathtaking cruelty. He was supposed to give his worst enemy the railroad he had devoted his life to building, for him to run however he liked? If Strong's initial request for thirty percent of the Rio Grande was like Strong's asking to borrow Queenie for the weekend, this was like his asking him to throw in his castle, town, and daughter, too.

Every fiber of Palmer's being must have recoiled at the notion, but he had no choice. His railroad was dying an agonizing death, squeezed by Strong's foot on its neck. The company's revenues were shrinking, its costs mounting. But the General came to see that, as bad as a lease was, it was not a sale. Leased, the Rio Grande would live on even if it was no longer fully his. And there was this, too: if Strong had the Rio Grande, he wouldn't kill it by paralleling its tracks.

The General had no choice but to accept Strong's offer—his share-holders would have rebelled otherwise—but the whole idea was so painful he couldn't bring himself to participate in the negotiations. He left that grim business to a pair of major Rio Grande bondholders, con-ducted at the Santa Fe's grand offices on Devonshire Street in Boston.

Strong also removed himself from the negotiations, leaving them to President Nickerson and his brother Joseph, a fellow board member in the manner of the Boston Crowd. Both Strong and the General watched the proceedings from afar, but the General far more atten-tively as he hunted for angles that might save him.

Sure enough, he found some. Officially, the lease would run for thirty years, a near lifetime, but the General thought that perhaps he could slip in some innocent-seeming provisions that might make it less. One was to require the Santa Fe to keep all the Rio Grande machinery in good working order. That seemed reasonable enough, and it was in the Santa Fe's interest, too, surely? The Nickersons agreed. That brought a knowing smile from the General, as he contemplated what might happen if, in his judgment, they failed to maintain the RG to his standards. Another was that the Santa Fe make its lease payments monthly. Perfectly understandable, and also agreed to. Another smile as he thought about the possibilities if the Santa Fe missed one.

Finally, Palmer asked the Santa Fe to allow the Rio Grande to build on to Leadville above the twenty-mile Dead-line—with the provision that the Santa Fe could then buy the road when it was completed for a price that covered the Rio Grande's construction costs. The Nickersons thought that was ridiculous. But, in rejecting the idea, they inadvertently excised only the part about the Santa Fe buying the line at cost—the part about allowing the Rio Grande to keep building remained a condition of the sale.

Even though the General had been kept fully apprised of the developments, and indeed was personally responsible for many of them, he professed shock to see the final agreement at the end of November 1878, nine months into the conflict. He was expected to sign immediately but refused. Instead, he set off for Boston determined to get better terms.

Before he arrived, a press-savvy Strong tipped the Boston newspapers that the deal might be off. The Rio Grande stock duly plunged at the news. When Strong hinted the deal might be on after all, the stock shot up again. Palmer got the message. He signed.

The deal was quickly ratified by the investors, leaving Palmer obliged to turn over his railroad to the Santa Fe on December 1. But the day passed without Palmer doing any such thing, as he claimed that Strong had scotched the deal when he questioned the Rio Grande's valuation of some of its construction equipment. Indignantly, he referred to it as an "arrogant demand of possession before complying with the plain terms" of the agreement. "I have declined, of course, point blank and expressed amazement at [Strong's] demand," the General declared. He thought he finally had Strong in a bind—unable to take possession of the RG or to leave it alone since so many members of the Boston Crowd had bought in. Lambasting Strong's questioning of the valuation as proof of bad faith, Palmer demanded the Santa Fe put up more financial security to ensure the proper handling of the Rio Grande's physical equipment—its trains, track, buildings. In even higher dudgeon he declared that no amount of security would be enough from Strong's Santa Fe. "[If] they were to put up Boston itself now, it would not avail," he snapped, professing disgust at the Santa Fe chiselers who were "insolent & arbitrary and more so at Boston recently." Things were, he concluded, "in a thoroughly antagonistic shape."

The genteel Thomas Nickerson couldn't believe it. "I did not think that after making peace we should still have war," he told friends. Palmer was standing on air, and it did not hold him up for long. Refusing to be drawn into any war of words, Strong simply reiterated that the Rio Grande still needed to deliver a reliable appraisal of the construction equipment it would be surrendering. Forced to acknowledge the fact that he had not furnished one, the General tasked Alexander Hunt with collecting the numbers—and then told him to make himself scarce in hopes of further delaying the transfer.

It didn't work. The Rio Grande shareholders were too eager to convert their sinking shares to rising Santa Fe stock as part of the lease agreement. They did not want the deal to blow up over the price of some shovels. Tired of the theatrics, they told Palmer to settle.

Backed into a corner, the General informed Nickerson that if he was wired a $150,000 guarantee toward the tool payment he'd lease the Rio Grande. Nickerson agreed, but, sick of Palmer's antics, he said he'd deliver the $150,000 in two installments—the first *after* the General surrendered the railroad and the second *after* he provided the long-sought valuations. Palmer had no choice but to accept these conditions. The deal was struck.

On Friday, December 13, Palmer invited Strong to Colorado Springs to convey to him the railroad he had built from scratch. If Strong was the new boy when the two had met a year before, he certainly was not now. Strong had run lines all over Colorado, outfoxed the General at every turn, and driven him to his knees. In abject defeat, Palmer might have made a show of his usual self-confidence, but he was hurting inside. Strong didn't much care. To him, this was just another deal.

The meeting likely occurred at the Colorado Springs telegraph office, since the lease would not go into effect until Palmer cabled instructions to subordinates, and it is easy to imagine how it went down. The deadline was midnight, so it occurred as darkness encircled both of them since Palmer would not have wanted to give up his railroad a second sooner than he had to. The meeting itself was little more than a handshake and a brief meeting of the eyes. Strong's hand felt bigger than ever when Palmer shook it.

To make sure Palmer delivered, Strong watched the telegraph operator key in the message to all the Rio Grande employees that William Barstow Strong, general manager and vice president of the Atchison, Topeka and Santa Fe Railroad, was officially the boss of all the employees of the General's Denver & Rio Grande, and he was in charge of the line everywhere it ran.

For now.

CHAPTER 14

This Means War

EVEN BEFORE STRONG WAS OUT THE DOOR OF THE TELE-graph office Palmer was already planning to take back his railroad. He directed Robert Weitbrec, his treasurer, not to accept any of the Santa Fe's monthly payments for the rental of the road as required by the lease. If they were not accepted, they had not been made, right? In Palmer's view, this would void the agreement, and restore ownership to him. He would also keep an eye out for the first scratch on one of his locomotives or passenger cars.

For his part, Strong had no interest in running the Rio Grande anywhere except into the ground. To him, the line was little more than a toy clattering about the Rockies to annoy him. Palmer might talk of springing to Mexico, or leaping over the mountains into Utah, but he had yet to expand out of the state. Strong would drain the Palmer's Rio Grande of everything it was worth and use the revenues to push his Santa Fe west to California.

Shortly after the lease agreement was signed, the Union Pacific joined with the Kansas Pacific to steer their traffic to Denver and away from Pueblo. Strong could see it was their way of bringing the newly frisky Santa Fe to heel.

How to fight back? By choking the life out of the Rio Grande. He drove up the rates on the Rio Grande trains linking Denver to any points south, thus drawing traffic back to the Santa Fe. If it drained the RG's revenues, so much the better. By the lease agreement, Palmer received one dollar out of every three on his line, money he desperately needed to pay off his bondholders. If they sued for payment as those Dutch investors did, he'd be responsible. Strong refused to honor any Rio Grande passes on the RG now that it had been leased to the Santa Fe, stranding any number of dignitaries—who then took their frustrations out on a powerless Palmer.

By now, Morley had completed the Santa Fe line from La Junta to Trinidad, and Robinson was almost done with the shoo-flies to hurl Santa Fe trains over the Raton Pass. With the former in service, Strong made sure that it carried out all the coal in the region, starving the Rio Grande's line to El Moro, and cutting so sharply into its coal business that the town soon disappeared entirely, taking revenues with it.

Strong's only lingering worry was the Supreme Court. Initially, he had been confident it would let Hallett's ruling stand, but in January, he got wind that the court might be leaning toward the Rio Grande after all. To prevent any shift, he swamped the Supreme Court with Santa Fe lawyers, ordering them to produce enough motions to keep the court from ruling until the Santa Fe had completed its line to Leadville. To be on the safe side, Strong also had them make a slight alteration to an obscure bill pertaining to the legal structure of the state's churches to allow a majority of corporate stockholders to overrule its directors' decisions. If the Supreme Court ruled for the Rio Grande, Strong could scotch any attempt by the RG board to take advantage

In the meantime, he kept building, flooding the gorge with a thousand workmen, all relying on Rio Grande trains to haul in the heavy rails for quick installation in the gorge. When Palmer sent McMurtrie to monitor Strong's efforts, the General was sickened to learn of his rival's progress. Unable to halt it, Palmer ran to the local Colorado Springs *Weekly Gazette* to decry the Santa Fe's legal "chicanery" and claim it had failed to pay its monthly fee (never mind that the Rio Grande had simply not accepted it) and damaged Rio Grande's rolling stock in numerous but unspecified and unsubstantiated ways. The charges found sympathy in Denver, which had been hurt by the high ticket prices Strong had imposed on the Rio Grande. The Denver *Times* went so far as to demand the state give Palmer back his train.

Strong fired back in the Santa Fe–friendly *Colorado Chieftain* that it was all poppycock. The Santa Fe had made every one of its payments— Palmer had simply failed to deposit them. And the last thing the Santa Fe wanted to do was damage the trains it had leased.

The newspaper war split the state between the two railroads, but this time the public sided more with the General. Strong's Santa Fe might once have been the railroad of the people standing up to the General's snobby Rio Grande, but now the Rio Grande was the home-grown darling fighting off the dastardly corporate invader. To the *Rocky Mountain News*, the Santa Fe was "the most grasping, blood-sucking Corporation Southern Colorado has ever known." A resident of Pueblo derided the once-loved Santa Fe as a "banana line" operating out of "luxurious offices in Boston" bent on "bleed[ing] out merchants and shippers of the last cent," in an attempt "to absorb the pioneer railroad of the state by subterfuge and trickery."

It seemed that Cañon City alone remained loyal to the Santa Fe. "The complaints of injustice at the hands of the Rio Grande in multifarious ways were widespread and very bitter," wrote one correspondent of the town's sentiments. "That Company is charged with all the offences of the calendar and credited with no virtues."

In March, just three months into the lease, Palmer sent Willie Bell to Boston to bring a suit against the Santa Fe in its home jurisdiction, demanding his railroad be restored to him for all the reasons he'd detailed in the *Weekly Gazette*. In April, Palmer went a step further, convincing the state attorney general to claim that since it was incorporated only in Kansas, the Santa Fe was operating illegally in Colorado. No matter that the AG had been a partner in the law firm Palmer hired to defend the Rio Grande.

Then, the General decided he'd had it. Confident as he was that the courts would see things his way, he was not going to wait for them. He'd decided to assemble a private army to take back his trains, tracks, depots, roundhouses, and maintenance shops all along the Rio Grande lines. While he was at it, he'd throw the Santa Fe work crews out of the route to Leadville, too. By now, DeRemer had seventeen stone forts all along there; he'd fill them all with well-armed Rio Grande soldiers and dare any Santa Fe workmen to show themselves

When he learned of the General's plan, Strong did not give an inch. He hired gun-toting soldiers of his own—and not just any soldiers. He recruited none other than his old friend Bat Masterson, now Dodge City marshal, to round up as many gun-happy cowboys as he could find and speed them all to Cañon City on the Santa Fe. From there, he'd sent them out along the Rio Grande tracks to protect the property he'd leased, and he directed them to go up from the gorge clear to Leadville. Let General Palmer cope with *that*.

As wars go, this was a strange one. It was like the initial confrontation across the gorge, where both sides were far more intent on a massive display of intimidation than actually shooting anyone. It was, if anything, an effort to avoid a fight, not to have one. Little of the ammunition that was furnished to either army was live; and any lethal bullets were discharged only to the sky.

Unlike in a military war, in a business dispute it was murder to kill someone. The law was still patchy in the West, but homicide was still a capital offense, and as likely to be enforced by a gang of vigilantes as by a formal court of law. What's more, since the conflict was largely being fought for public opinion, it would not help the cause for one side to start mowing down the opposition. And it was a conflict oddly without impact. The confrontation was spread over the Rio Grande lines throughout the whole state of Colorado, making the threat to property in any one locale, even Denver, negligible. There was no equivalent yet of a wealthy industrial class that was likely to feel endangered by the disorder—if anything, the industrial class in Colorado was largely comprised of the officers and representatives of the two railroads in the fray, the Rio Grande especially. If they sought to minimize the damage, they could make peace. If they didn't, they had no one to blame but themselves.

Nonetheless, the whole business looked scary. As the armies on these two sides grew larger in number and closer in proximity, tensions rose wherever they were pressed against each other. A sharp word could lead to a scuffle, and a scuffle to gunfire, and once bullets started flying, who knew where some live ammo might lodge.

The days ticked by in mounting suspense, neither side willing to budge, until April 21, when the United States Supreme Court finally ruled on the question of which railroad company held the rights to the gorge. The answer was—*the Rio Grande!* To the surprise of nearly everyone except Palmer, the court declared it did not matter that the Santa Fe was first to start laying track in the gorge. The Rio Grande held the rights by virtue of its initial plat.

In Palmer's Colorado Springs, joyous crowds gathered everywhere to celebrate, a few exuberant residents blowing the steam whistle at the sawmill until it had no steam left. But enthusiasm waned when it became clear that the decision of the court did not change the facts on the ground. While the ruling clearly favored the Rio Grande, it did not grant the company an exclusive right to the gorge. The Santa Fe could keep building. Now, the Rio Grande had first dibs, but the Santa Fe was still allowed to build wherever its rival didn't. As before, the two lines would have to share where there was room for only one set of tracks. Past the twentieth mile Dead-line, the original rules remained in effect.

What the decision did not address was the fact that, by now, the Santa Fe had completed its tracks through the gorge to the Dead-line. Was it supposed to just hand them all over? And there was a far bigger issue that had risen up since the litigation began: Now, in effect, there was no Rio Grande. It had been leased in its entirety to the Santa Fe. Was that lease no longer operational? Since it was outside the scope of the litigation, the Supreme Court decision offered no opinion, but, even if it had, no ruling could reverse the fundamental economics that had forced Palmer to give up his company in the first place.

Nonetheless, the ruling lifted the financial prospects of the Rio Grande, for it allowed Palmer to raise money off this unexpected good news. He rushed East to market his victory to eastern investors and hauled in $10 million. There was life in the old dog yet.

The ruling put Strong, however, in a rare bind. He had sold a million dollars' worth of bonds and even more of stock on the cheerful assumption that the company had the Royal Gorge all to itself. The *New-York Tribune* observed with savage delicacy that the court decision "promises to raise a multiplicity of nice points" for the Santa Fe if the ruling ended up bringing the Rio Grande back to life as an independent entity.

Strong was not one to give in, especially now that the value of a Leadville exclusive was growing by the day. A Congressional act restoring silver to the currency was fully in effect, and the Silver King Haw Tabor was achieving national renown. After splashy eastern publications like *Frank Leslie's Illustrated Newspaper* trumpeted the astonishing news of the town's silver riches, everyone wanted to get in. A hundred newcomers a day were racing up to two-mile high Leadville, clogging the streets on their way to a fortune. "The noisiest place you can imagine," said one new arrival. "You possibly have no idea of the rapidity of the action here," wrote another who'd taken refuge in one of Leadville's many saloons. "All is push and bustle in the shadow of seven bottles of some odoriferous substance." As more miners came in, more silver went out. Silver production jumped from $2 million in 1878 to $9.5 million in 1880 and rose from there.

Of all the facts on the ground, a key one was this: Santa Fe tracks had reached DeRemer's Dead-line and were poised to ascend to Leadville from there. No court had blocked them from going farther. So the General was determined to do that himself. Judge Hallett may have ordered all gunmen out of the gorge, but the General would use troops to keep them from coming back. He armed fifty of his workmen with shotguns, rifles, pistols, and bayonets, and deployed most of them in the big stone fort that DeRemer had erected to guard the Dead-line. He spread the rest farther up the route.

Seeing such a display of force, Robinson prudently stayed on his

The Rio Grande "soldiers" guarding the Dead-line.

side of the Dead-line. But at Strong's direction, he stockpiled all the materials needed to lay track the whole way up to Leadville, leaving no doubt about his intentions.

For his part, the General instructed his men to stay right where they were. The Dead-line was his line in the sand, marking the boundary of his domain. Though the highest court in the land had taken his side, no court could reverse time and turn life back to what it was when Palmer filed his appeal. Given that, he had to wonder—what difference did the law make, anyway?

⊱•◦•⊰

It was a fair question. At the end of 1878, Colorado had been a state for only two years. Most of its western neighbors were still territories and would remain so for decades more. While the Colorado governor held the power to call out the state militia, it was a largely untrained force of irregular volunteers. There was no police force worth the name, just city marshals with a few deputies, who concerned themselves with individual crimes like murder, fraud, and theft. Horse theft was still on the books as a hanging offense, and, in mining camps, a five-dollar theft was enough to earn a noose from Judge Lynch. There was no state police, let alone any FBI, to deal with the larger-scale crimes of more powerful interests. In 1864, an innovative Denver city marshal named Dave Cook had the idea of creating a regional police force, the Rocky Mountain Detective Association, to counter broader-scale criminality, relying on cable communication to coordinate crime fighting across the West from Wyoming to Texas. At its height, it consisted of over a hundred cowboy detectives, most of them city marshals, and accounted for several thousand arrests over the thirty-five years of its existence.

But even that effort was somewhat ad hoc, designed to solve only the crimes for which there was reward money. While the Wild West was often thought to be populated by murderers and desperadoes, such criminality was mostly confined to seedy hotbeds like Deadwood, Tucson, and Dodge City that were filled with drunken cowboys out for a good time. Elsewhere, life was fairly sedate; people needed to be good neighbors to survive.

In the territories, and in fledgling states like Colorado, government was not designed to serve voters so much as the powerful moneyed interests who controlled the fortunes of the elected officials. The railroad men were at the top of this list, but cattlemen, developers, miners, and wholesalers had plenty of say. When those interests were threatened from below by, say, a miners' strike, the governor might dispatch the militia to preserve order. But discord was much harder to contain when two powerful interests clashed, for each could usually call on friends in government to take their side, making the conflict nearly impossible to resolve.

When the Rio Grande and the Santa Fe faced each other down over that Dead-line at the top of the gorge, another such war had broken out in New Mexico's Lincoln County, pitting two rival factions against each other as they battled for monopoly control over a local dry-goods business. On one side stood "the House"—rich businessmen who supported the owner of the area's original dry-goods store. On the other, "the Regulators"—wealthy cattlemen who backed an English newcomer who sought to start a store of his own. This might have remained a local disagreement, but the House recruited to its side the Lincoln County sheriff along with a gang of outlaws. The Regulators enlisted the help of the US marshal and a variety of killers including Billy the Kid, and when a House gunman killed the Regulators' English storekeeper, all hell broke loose. Each murder was avenged by another for the next three years. As the body count mounted, each side co-opted enough of the government that the New Mexico territorial governor was paralyzed. The whole bloody business climaxed in a five-day shootout that left a dozen dead and would have gone on forever if the US Army hadn't decided enough was enough, and wheeled in cannon to scatter the combatants, bringing peace at last to Lincoln County.

As in Lincoln County, so it was in the Royal Gorge. Where there is no law, there is only force. Once that became obvious to the General, he abandoned any Quaker pretentions. This meant war.

CHAPTER 15

Bad Men

IN PREPARATION FOR BATTLE, THE GENERAL MADE GLEN Eyrie his headquarters. He had his men dig evacuation tunnels out from the basement, should Strong's men try to storm the castle and take him prisoner. He also ordered McMurtrie to South Pueblo, near the center of the state, to set up camp as Palmer's field general. He communicated with him by coded cable, making sure he had enough conscripts and arms. Palmer relied almost entirely on a cipher system that replaced key words with ones from an ever-changing codebook. "NEW PAGE FLESH AFFABLE CALIBER," ran one Palmer message. "NO BEMOAN WAD TEMPEST." Translation: "New pistols forty-four caliber. No ammunition with them." However surreal in code, the messages were dead serious. When McMurtrie told Palmer he had a full box of Springfield rifles and thirteen Colt pistols to distribute, the General replied that wasn't nearly enough, and sent down thirty pistols and twenty carbines, plus live ammunition by wagon.

Gun tallies came in to Palmer from all about the state. His man in El Moro reported he'd handed out six rifles and six pistols to Rio Grande soldiers and had plenty more on hand: six Colt revolvers "and one box of ten rifles, and another of ten pistols had just been delivered." At another Rio Grande stop, his forces had ten rifles and six revolvers at the ready. Palmer carefully recorded each total back in his office in his tidy hand. By January, his troops had 207 pistols and 259 rifles altogether.

With McMurtrie installed in South Pueblo, Palmer sent Willie

Bell—back now from his legal pursuits in Boston—taking charge else-where in the state. On June 8, Bell reported from Cañon City that he had Rio Grande soldiers in position at nearby Silver Cliff, and that he'd sent yet more rifles to bolster the forces at El Moro. He also passed on a key bit of intelligence: The El Moro telegraph operator was secretly partial to the Santa Fe. For days he'd been destroying Rio Grande mes-sages without sending them, but Bell had fixed that. He'd sent a couple of Rio Grande men to pull him from his post and hold him captive.

With the time to strike approaching, Bell recruited El Moro miners loyal to the Rio Grande to board trains for deployment in the north. "A splendid body of 16 mounted men came in tonight," he told Palmer from Cañon City. "4 more come tomorrow." He closed by assuring Palmer he'd been careful to avoid detection as he roved about, as he was fully aware there were Santa Fe spies everywhere. "The enemy watch our every movement & send armed men after DeRemer & my-self to see what we're about."

Strong—now relocated to Denver, closer to the action—was also gird-ing for war. Details are spotty, but records of the company show his expenditures were commensurate with Palmer's: "Colts Revolvers and Ammunition, $112.30"; "Arms and ammunition in April, 1879, $94.50"; and "Payroll for May, W. R. Morley's Gang, $6,018."

Tensions rose across the state as key locations filled up with soldiers from both sides. In Pueblo, rough-looking men knotted in the streets or caroused in taverns as they awaited word from higher-ups about how to proceed. All of them had come from somewhere else, and were, by one local account, "staggering about heavily laden with blankets and traps, swearing, yelling, and making themselves felt as forceful masters of the situation." At any moment these "bad men" could fly by train to bring their war to almost anywhere.

Forceful as they were, masters of the situation these combatants decidedly were not, with the possible exception of Strong and the General—but even they were far more in the grip of each other than in command of themselves. A war was on, that much was clear. But who knew for what? Did the General really think he could reclaim his railroad by force and set it back into operation under his own control, reversing law and economics? Did Strong really think that a fight might secure his rights? Would either man actually direct his troops to shoot to kill? It all made for a cruel reversal to the state's notion of progress. No longer the benign cultivators of the prairie, the Rio Grande trains were returning it to a howling wilderness. On a personal level, Strong and the General had long since lost track of who they were independently of each other. And, even if they had clung to their own singular characters, they were themselves knocked about by economic, political, and judicial forces far beyond their control.

Nothing quite illustrated the capricious nature of it all than the lease itself. Who possessed it? That piece of paper, with all its legal flourishes, represented all there was of the Rio Grande railroad—its track, rolling stock, employees, everything. While the lease had been surrendered by Palmer, it was not held by Strong, Nickerson, or anyone at the Santa Fe. It was instead, like the railroad itself, suspended in a curious betwixt-and-between state, in the possession of Sebastian Schlesinger, one of the two bondholders who handled the original Rio Grande's negotiations with the Santa Fe. An obscure Boston financier known for little beyond a fondness for classical music, he was one of the two largest owners of Rio Grande bonds, and so had become the escrow officer for the transaction, and as such, the titular holder of the Rio Grande. But those bonds were now held by the Santa Fe as the lessor of the rival line. Thus Schlesinger stood with a leg on both sides of the deal, representing neither party.

A further wrinkle: Schlesinger was also the first husband of the unhappy but highly pedigreed Countess Berthe de Pourtalès, the woman who had so fascinated Frances Wolcott when she came to Colorado Springs with her brother. Count Louis Otto de Pourtalès was the "Mute Seraph" and "born lover" in Wolcott's account who may have been practicing his skills with Mrs. Palmer in her third-floor bedchamber, putting Queen in a state of play not unlike her husband's railroad—possessed by him as her husband, but, in effect, borrowed by Louis.

This whole notion of "possession." What did it actually signify? It looked like the Indians were right—no one could actually possess something of such value to so many. The best anyone could do was use it and keep others from it by force. And what kind of a "war" was this? For all the gnashed teeth, mobilization of troops, and distribution of arms, not a shot had been fired. Was this, like all the skirmishes that had come before, for show only?

That seemed to be the case until the beginning of June. A Judge Bowen had been enlisted to rule on the question of the Santa Fe's corporate status raised by Colorado's compromised attorney general, C. W. Wright. This was a decision Strong had not given much thought to, since it seemed so preposterous that, in a big country, a company had to be incorporated everywhere it did business. It turned out he should have. To Strong's horror, Judge Bowen ruled in favor of the Rio Grande. Since the Santa Fe was headquartered in Kansas City, it had no lawful right to operate in Colorado. It needed to cease all its operations in the state immediately.

Thunderstruck, Strong gave in to Palmeresque outrage, unleashing a blistering personal attack on the judge. Nonetheless, Bowen issued a court order forcing the Santa Fe to give the Rio Grande back to Palmer. In a panic, Strong rushed lawyers to appeal to Judge Hallett, who had issued the original favorable ruling about the Santa Fe's right to the gorge, in hopes of overturning what the ever-loyal *Colorado Chieftain* termed "Bowen's Infamy." In the meantime, no action could be taken to enforce Bowen's order. The county clerk—in sole possession of the official seal that would formalize the order—had mysteriously disappeared. Rumors soon flew that Santa Fe henchmen were plotting to kidnap Bowen as well to keep him from officially issuing his decree.

Hunkered down in Colorado Springs, Palmer was ready to believe the worst about the Santa Fe. Strong sent his men to cut down enough telegraph poles to sever Palmer's cable connections to the entire southern part of the state. Palmer begged the new governor Frederick Walker

Pitkin in Denver to restore the telegraph network before Colorado descended into chaos. To enlist popular support, his lawyers instructed the Rio Grande's general manager to tell the Denver newspapers to "square up to the governor. Give it to him right between the eyes." Governor Pitkin replied that there was little he could do about the telegraph lines, but he would send in state militia to back up any local sheriffs enforcing Bowen's order.

The General, however, knew that the state militia would not be enough. He resorted to bolstering their forces with gangs of his own well-armed men, effectively making the militia an advance guard of his private army. He insisted that the sheriffs immediately move up and down the Rio Grande lines to act on Judge Bowen's order and return to Palmer his railroad before a court could say otherwise. They were to fan out throughout the state to every depot, roundhouse, train, and stretch of track, boot out the Santa Fe people, and reinstall his Rio Grande employees. Palmer would be the General once more. He would take his Rio Grande back or die trying.

Anticipating his opponent's fury, Strong hurried troops of his own to the Rio Grande properties to hold Palmer's men off. To secure Pueblo, the linchpin of Colorado's two railroads in the middle of the state, he rushed in three trainloads of soldiers, forty-five of them from Trinidad under the leadership of an Irishman named Paddy Walsh; eighteen more under the direction of an SF-friendly sheriff; and sixty-five wild-eyed gun-toting cowboys from Dodge City led by Bat Masterson.

The General struck first. At 6:30 a.m. June 11, he targeted his hometown of Colorado Springs. There, he had the local sheriff, backed by a dozen deputies and Company B of the First Colorado Cavalry, stride up to the Rio Grande depot to serve Bowen's order to any Santa Fe officials inside. Strong had stocked the building with a dozen well-armed men for just this moment. Not wanting to risk a bloody confrontation, the sheriff spotted another Santa Fe operative outside, and served the warrant there, rather than take on the men inside. Then he returned to the depot to shout through the door that the deed was done and any further resistance was pointless. "I don't want any of your foolishness," he yelled. "Open that door and come out or I'll break it down." Unwilling to risk their lives to defend a depot that was no longer theirs, the men put down their guns and stepped out. The depot at Colorado Springs was the Rio Grande's once more.

Similar encounters occurred all over the state. In Denver, the sheriff was forced to burst down the door of the railroad offices to serve his writ, but elsewhere, the Santa Fe forces bowed to judicial authority, backed by Rio Grande soldiers, without much resistance. Alexander Hunt brought a force of two hundred up from the south, liberating stations as they went, without bothering with any judicial paperwork. They fell like dominoes as Hunt hurried along on Palmer's behalf.

It was in Pueblo that Strong made his stand. McMurtrie sent out the sheriff, accompanied by a crew of Rio Grande thugs, to extricate the Santa Fe men from the train dispatcher's office, but this time, the Santa Fe soldiers refused to budge. What the papers called a "shooting fray" ensued, though the guns were fired mostly in the air. Realizing how many men Strong had collected in Pueblo, Palmer summoned reinforcements, who arrived late in the evening. By then, the Santa Fe had concentrated its men inside a brick roundhouse, large enough for a locomotive to turn around inside. Bat Masterson had filled it with his Dodge City rowdies, bearing instructions from Strong to hold the building at all costs.

The many florid newspaper accounts differ as to what exactly happened next. In one, Masterson refused to open up when the sheriff pounded on the door, and so the Rio Grande forces seized a militia cannon to blow the door down and storm inside; the Santa Fe's men surrendered after Masterson jumped out the window. In another, the Rio Grande treasurer banged on the door, informing Masterson that he had so many men with him he'd better open up. Masterson complied without a fight. A third account had a Rio Grande representative offering Masterson $25,000 to surrender the roundhouse. When he refused, Rio Grande men took it by force. A fourth claimed Masterson was not even there, and the seizure of the roundhouse by the Rio Grande was a nonevent. The variety of the accounts reflected the chaos provoked by fierce partisanship in the state, which left few impartial observers to provide accurate information. As a result, the newspapers followed their own prejudices. In any case, the Rio Grande retook the round-house by nightfall. Palmer had his railroad back.

To meet his men, the General boarded a train to Pueblo about that time. On it, he spied another Rio Grande train coming up toward him on the same tracks. Was it carrying friends or foes? Both trains stopped, and Palmer cautiously stepped out, pistol in hand, unsure who he was about to encounter. A few passengers on the other train did the same, approaching him with guns drawn. Finally, Palmer recognized the other men as his own, and they recognized him as theirs, and as one man put it, "peace prevailed."

The only question now was, how to maintain it? For Strong to re-claim the Rio Grande, he'd have to get a court order of his own—not an impossibility given that his lease was still in effect. For the General, the only thing to do was the very thing he had been desperate to avoid. He had to put his railroad into bankruptcy. That way he could choose a friend to take charge of its operation as the receiver. The gambit worked. When Strong sent lawyers racing to Judge Hallett to get the Rio Grande back, he discovered that was legally impossible. The General could not give what he did not have. It was in the hands of the receiver now.

When another judge in the building heard of Palmer's maneuver, he exploded, "No judge, no court can sit quietly down and tolerate such abuses of process." To that, wrote one newsman, the Rio Grande lawyers responded "as if they had been struck by lightning," but they did change their legal strategy.

In the end, Judge Hallett did offer a refinement to the Supreme Court ruling. Palmer could retain his priority all through the gorge and all the way up to Leadville, but, if he wanted to take the Santa Fe's tracks to the Dead-line, he'd have to pay for them in full. Hallett ac-knowledged that it would likely take a while to settle on a price, since Strong did not want to sell, but in the meantime, Palmer could con-tinue building to Leadville. The Rio Grande would stay in bankruptcy, but not under Palmer's receiver. Hallett would choose a new one who had no loyalty to either railroad company.

Strong was so furious, he declared he'd kill off the Rio Grande after all. He'd shadow the Rio Grande's lines throughout the state and squeeze the life out of it.

⊱─◦─◦─◦─⊰

At the beginning of 1880, the war for Leadville had been raging for nearly two years, and yet it was no closer to resolution than the day it began. Childish to begin with, it had long since become a total embarrassment for all concerned. Even when Palmer won two definitive legal victories, enlisted an army to enforce them, and took back his railroad by force, he still didn't possess his Rio Grande, nor had he gotten any closer to securing Leadville. For him, the problem wasn't the law. It was reality. The numbers were against him. He could not alter the fundamental economics that had put him at such a gross disadvantage to Strong that he had to lease him his railroad in the first place—and despite his financial advantage, Strong was no better off after his two bitter legal defeats. He couldn't exercise his right to Palmer's train, and he couldn't claim Leadville, either. For Strong, the problem was metaphysical. The whole controversy had become engulfed in so many contradictory court decisions, distorted newspaper accounts, rampaging armies, inflamed editorials, and general uproar, that everyone's head was spinning.

Most observers tended to chalk up this whole fiasco to the competitive juices of the extremely ambitious. That made sense, but it overlooked the fact that the two men persisted in the conflict long after it was likely to ruin them both. The solution was right in front of them: share the route, and then share the rewards (or at least negotiate mutually beneficial terms). No winner, no loser. Equals. But neither man considered it. Instead, they fought on, making everything worse for both of them every day.

Why? Some of the fury of the contest might have been drawn just as much from the men's private lives. On Palmer's side, he had his mounting frustration with his Queen, who was denying him. On Strong's side, it is hard to tell since so little of his personal life is known, but he had to answer to corporate bosses who were far removed from the fray.

As ever, capitalism played a major role. Railroad men, not to mention silver prospectors, real estate developers, dry goods entrepreneurs, cattle rustlers, and most of the other more energetic actors in the economic life of the West in this period, hungered for an exclusive, something they could take entirely for themselves, even if it was the size of California. Usually, such scrambles for a prize are a free-for-all, one among many, defusing some of the ire, but Strong and the General were competing solely against each other. That personalized it.

The force behind their railroad war wasn't just blind rage but the targeted version, a jealous hatred that threatened to destroy these two men, and their railroads, too. Strong and the General were like two mortal enemies grappling on the edge of a cliff, each with his hands tight around the throat of the other. Neither could let go, but neither could finish off the other, and both were in danger of toppling into the abyss. And no higher authority could get them to stop, for there was no higher authority. Not the public, not the courts, not the government, not the law.

And then there was. A god came on the scene. Seeing him, the two bitter enemies were so awed they relaxed their death grip and turned in fear toward this divinity who had the power to annihilate them both.

CHAPTER 16

Gould

BY 1880, JAY GOULD WAS THE REIGNING LORD OF THE
railroad world, and nearly everyone in it did his bidding. If he didn't
buy or sell railroads at prices of his design, he forced rivals to feed his
lines or steer well clear. While Gould saw himself as an investor, to
most everyone else he was a horror. Just five feet tall, withered from
tuberculosis, whispery voiced, Gould was to many a man of midnight—
black bearded, raven dressed, shadowy. "One of the most sinister figures
that ever flitted batlike across the vision of the American people," the
newspaper publisher Joseph Pulitzer called him. "A spider," said Henry
Adams, "[who] spun huge webs in corners in the dark." "His touch is
death," added a rival, Daniel Drew.

But if Gould was mendacious, his mendacity was, in the main, perfectly legal. If others scorned his methods, it might have been from
jealousy that he played their game so much better than they did.
Gould himself voiced no opinion on the subject. He wielded silence
like a blade. In company, he preferred not to speak, but instead ripped
a sheet of paper into small pieces, and then ripped those into smaller
pieces still.

Like so many other ruthless capitalists of the day, he'd come from
nothing. Born on a farm in New York's Catskill Mountains, Gould was
the youngest of six children, and the sickliest. His childhood diseases
were precursors of the debilitating tuberculosis that was to come. Useless behind a plow, he went away to school at thirteen, taught himself

An 1882 cartoon showing Jay Gould
bowling over the opposition.

surveying, and was soon teaching it to others. His own work led him
to a share of a tannery in the Poconos, which he built up to the size of
small town, dubbed "Gouldsboro." To buy out the majority owner, he
took on a couple of partners. It said a lot about Gould that one of them
imagined a vampire was perched on his shoulder and shot himself in the
heart. The other, convinced that Gould was cheating him, recruited a
handful of armed men to storm the tannery and barricade themselves
inside. Nonplussed, Gould rounded up a pair of assault teams, one to
attack the front door while the other stormed in the back, both of them
with guns blazing.

Gould took the tannery back, but the uproar rendered it worthless.
He left for New York City, married the homely daughter of a prosperous
Manhattan merchant, and expended some of his father-in-law's fortune

to win control of a tiny railroad that ran from Vermont to the Hudson River. A year and a half later, he added another line. Then he got ambitious.

There had been a stock exchange in New York since the founding of the republic; the original traders signed the agreement organizing securities trading under a buttonwood tree at 68 Wall Street in 1792. As late as the 1830s, after the traders organized a formal New York Stock Exchange, it was a rare day if more than thirty shares of stock changed hands. It made for a clubby affair. Members of the exchange wore top hats and prided themselves on their civility. But by 1850 the number of transactions reached thirty thousand, and the action spilled outside to the sidewalks where "curb brokers" could get in on the buying and selling of securities, introducing the distinction between relatively ignorant outsiders and knowing insiders who towered far above them. One Wall Street trader came up with a name for those on top:

These men are the Nimrods, the mighty hunters of the stock market; they are the large pike in a pond peopled by a smaller scaly tribe. They are the holders of those vast blocks of stock, the cubical contents whereof can be measured by an arithmetic peculiar to themselves; they are the makers of pools large enough to swallow up a thousand individual fortunes. Sooner or later, the money of the smaller tribe of speculators finds its way into the pockets of these financial giants.

These Nimrods soon became better known as the "Robber Barons," some of the richest men not just in the world but in history. If the term was meant to convey disdain, it did not to Gould. He was determined not just to join their ranks but to outdo them all.

He started small. As a sideline to the railroad business, Gould dabbled in the stock market through a brokerage house that won attention because of some stock it held in the Erie Railroad. At that point, the Erie linked greater New York City to the Great Lakes, but it ran from the forgettable Jersey City to Dunkirk, a spot southwest of Buffalo on Lake Erie, and skipped the heavy population centers in between. The tycoon Cornelius Vanderbilt was the Erie president, and he thought it would make a valuable addition to his New York Central. But his onetime friend Daniel Drew was the Erie treasurer, and he thought it might be fun to deny Vanderbilt the acquisition. The two of them were uncommonly unpleasant men. In his epic rise from steamships to railroads, Vanderbilt had once placed his wife in an insane asylum so he could more freely indulge his affair with the governess, and Drew had started out life as a cattle-drover herding cows down Broadway. He fed the cows salt so they'd gorge themselves on water, boosting their weight, and thus their price. It was widely said that Drew "watered" his stocks similarly. Unlike most people, Gould did not shy away from a fight between two these two vipers.

Realizing that Gould controlled some critical shares of the Erie, Drew sought to draw him into a scheme that he'd cooked up, and Gould was only too happy to oblige him—in exchange for a seat on the Erie board. Drew knew something about Vanderbilt that others overlooked. This tycoon was not rich in money, only in railroads, so he'd have to borrow the money to buy the Erie shares he needed to get control—$10 million worth, or $150 million today. To get them, he'd have to put up his existing Erie stock as collateral, and that stock would have to hold its price. But what if it didn't? What if those shares were watered down by, say, flooding the market with them? Drew could do just that as treasurer by issuing "convertible" Erie bonds that turned into stock whenever Drew wanted—namely when Vanderbilt relied on them to make his purchases.

At that point, Vanderbilt found he had to put up more and more of his existing shares in collateral to buy more shares for his purchase. It was a nightmare! Vanderbilt was trying to climb stairs that were dropping under his weight. The more cash he tried to raise, the deeper into debt he plunged.

In desperation, Vanderbilt persuaded a crooked judge, the Tammany man George Barnard, to forbid this "convertible" bond nonsense, and arrest the men responsible. Drew fled to New Jersey, but Gould coolly repaired to Delmonico's for dinner with his new friend "Jubilee

Jim" Fisk, a fellow Erie board member and high roller he'd drawn into Drew's scheme. Fisk brought along $7 million in cash, the Erie's entire treasury, in a satchel. They'd hardly finished their entrees when they got word that city marshals were converging on them. They departed without dessert, Fisk with his satchel, and paid a pair of sailors to row them across the Hudson. Vanderbilt hired armed thugs to kidnap them and return them to Manhattan where they could be tossed in jail, but Gould hired more than a dozen New Jersey cops to stand guard around his hotel. Drew bailed, but not before withdrawing the Erie's $7 million that Fisk had by now deposited in a New Jersey bank. When Gould sued to recover it, Drew surrendered the money and begged Vanderbilt to leave him out of this mess. Meanwhile, Gould snuck up to Albany to bribe enough legislators to restore the legality of Drew's "convertible" scheme and then paid Judge Barnard to rescind his arrest order. That did it for Vanderbilt. He could see he had met his match. He sold all his Erie stock and quit the board. Gould ascended to the Erie presidency, and made himself its treasurer, as well. At thirty-two years old, he was a Robber Baron.

Gould went on try to corner the national silver market in order to raise commodity prices for his freight hauls on the Erie but was foiled when President Ulysses S. Grant had his treasury secretary buy enough silver—which then backed the dollar along with gold—to drive the value of a dollar back down again. Fourteen brokerages, several banks, and countless individual investors were ruined by the currency fluctuation, and Gould himself lost millions. This is when he became a modern-day Mephistopheles, "the most cold-blooded corruptionist, spoliator, and financial pirate of his time" as one contemporary chronicler put it.

Before long, Gould came to relish his image as the devil incarnate, since it could be used to intimidate his business opponents. At the time, he retreated into his sumptuous private offices he called Castle Erie above the Pike's Opera House on Twenty-Third Street, where he was protected by a small battalion of armed guards from any disgruntled investors. He huddled there, nothing but "a heap of clothes and a pair of eyes," said his friend Fisk, who'd joined him in the silver misadventure.

Gould rebounded, but not Fisk. He was shot to death by his business partner, the hot-blooded Ned Stokes, after Fisk seduced Stokes's sultry paramour, Josie Mansfield. "It won't do, Josie," Stokes had pleaded with Mansfield after he learned about Fisk. "You can't run two engines on the same tracks in contrary directions at the same time." Sure enough, the two collided when Stokes greeted Fisk at New York's Grand Central Hotel with a bullet to his guts. It was the society murder of the century.

Gould decided to boost the position of the Erie Railroad he now controlled by taking a piece of Cornelius Vanderbilt's New York Central. To grab the cattle Vanderbilt carried east from Buffalo into New York City, Gould undercut Vanderbilt's ticket price of $125 per carload by fifty dollars. When Vanderbilt dropped his rate to fifty, Gould put his at twenty-five, only to have Vanderbilt go all the way down to one dollar, plus one cent for an individual hog or a sheep. Vanderbilt was overjoyed to see his cars stuffed with livestock while Gould's ran empty—until Gould bought up all the marketable animals in Buffalo and shipped them for a penny apiece on the Central. When Vanderbilt realized he'd been taken again, a Gould ally recalled, "he very nearly lost his reason."

━━●━○━●━━

Gould ended up being ousted from the Erie in a coup—only to score millions when his Erie stock soared on the news that the devil had departed. He used his winnings to take over the scandal-riddled Union Pacific in 1874 and threw himself into reviving it. He toured the whole line in his finely appointed private car, the Convoy, a stenographer beside him to record the many improvements he sought to make. When he purchased the Pacific Mail, a steamship line to connect the UP to the Far East, he became the man to watch on Wall Street. Wrote *Railway World*: "To write of the New York Market is simply to describe the movements of Jay Gould."

To protect his investment, Gould also bought into the UP's main east-west rival, General Palmer's old Denver-bound Kansas Pacific. In control of two western trains, he found himself troubled by a third, an ambitious line heading west at a lower latitude out of Kansas, the Santa Fe. And, come 1878, he could see that its ambitious new general manager, William Barstow Strong, bore watching. He could change the game considerably.

Strong's squabble with the General for Leadville had been a checkers match for two, but now, with Gould in the fray, it became four-dimensional chess for three. Of them, only one was a grand master. To hem in the Santa Fe from the West, Gould bought the Denver Pacific that linked the UP to Denver from the north. But then he feared that might send Strong into the arms of the Chicago, Burlington and Quincy, another Boston-based railroad that was starting to haul ore out of places like Leadville to smelters in St. Louis, to the detriment of Gould's transcontinental lines.

And there was a personal angle as well. Gould detested the Boston Crowd. The very proper Bostonian Charles Perkins, president of the Chicago, Burlington, led the pack. He declared that Boston exerted on Gould "the same disagreeable influence upon him that holy water [was] said to have to his great prototype." In a word, Satan.

This made Gould all the more determined to take the Rio Grande's side against the Santa Fe. In September 1879, after the Supreme Court ruled in the Rio Grande's favor, he bought up half the Rio Grande's voting trust certificates on the open market and added a cash outlay of $400,000 to reduce the railroad's crippling debt. This won him a seat on the Rio Grande board, and the leverage to tell Palmer to avoid making any deals with Strong for fear an alliance might threaten the Union Pacific and Kansas Pacific only a little less than a union with the Chicago, Burlington.

Understanding that Gould could hurt his chance of running up the gorge into Leadville, Strong lunged for another route up from the other side of the mountains owned by a new railroad, the South Park. But then Gould expressed just enough interest in the South Park to push the stock price out of Strong's reach.

With Strong focused on Leadville, Gould hit him at the other end of his line, back in Kansas, the source of his revenues. He threatened to do to Strong exactly what Strong had threatened to do to Palmer: double his tracks through Kansas and halve his income.

He decided to call the new line the Pueblo and St. Louis. (The name alone had to hurt.) It would run from the Santa Fe's own Pueblo through Kansas all the way to St. Louis. It would be the death of the Santa Fe—and then, in an evil flourish, Gould declared he'd hire Palmer's construction company to build it.

For Strong, it was sickening. Matters were made even worse when Judge Hallett revealed how he proposed to enforce Palmer's primacy in the first twenty miles up to the famous Dead-line: Strong would have to give the General all the track he'd built for free.

It looked like the Rio Grande had won the war for Leadville—save for one little thing. The Rio Grande was not in Palmer's possession. It was in receivership, waiting for Palmer to pay off its massive debts, and he lacked the cash for that.

But Gould had no shortage of money. The Rio Grande could win the war for Leadville if he funded it.

So it was that Jay Gould held in his hands the fates of the Rio Grande and the Santa Fe. It's doubtful he enjoyed his triumph. For him, the chief satisfaction in such victories lay solely in the leverage he gained from them for greater conflicts. Since he had maximum leverage over both these lines, capable of annihilating either or both as he pleased, he could win from them whatever resolution he sought. So, when Gould summoned Strong and the General to Boston to make peace, they were in no position to resist—or, once they arrived, to quibble with Gould's terms.

They were the following:

No more litigation. This agreement is binding and final.

The Santa Fe relinquishes its lease of the Rio Grande and returns the railroad to Palmer, gratis.

The Rio Grande pays the Santa Fe $1.4 million for its track through the gorge, plus $400,000 for expenses.

Gould drops his plan to build the Pueblo and St. Louis.

To avoid future conflict, the two railroads are to stay away from each other for ten years. The Rio Grande is permitted to build only north of Pueblo, and the Santa Fe only south of Pueblo. Thus, Rio Grande receives the exclusive right to the route up the gorge to Leadville, and the Santa Fe the exclusive right to go over Raton into the Southwest.

That was it.

As simple solutions to complicated problems go, this one was ingenious. Essentially, it drew a line east-west through Pueblo, and left the Rio Grande to its north and the Santa Fe to its south, the two not to trouble each other again for ten years. The General got the exclusive to the Royal Gorge and the riches of Leadville above, while Strong received the exclusive to the Raton Pass and the vastness of the Southwest below. Both of them got peace.

What Gould got out of it was harder to discern, but ultimately more valuable. By keeping the pesky Santa Fe south of Pueblo, he made sure it would not venture into the territory of his Kansas Pacific to the north, or of his Union Pacific beyond that. The Santa Fe would not trouble him for a decade, an eternity in railroading. The ambitious Strong would be expending his energies going away from Gould, not toward. And now that the Robber Baron had control of the Rio Grande, it wasn't going to bother him to the north.

For the moment, General Palmer looked like the clear winner, as he now had exclusive claim to the riches of Leadville the two had been fighting for. The Santa Fe railroad was left only with the desert of the Southwest. To be sure, that might open up a path for the Santa Fe railroad to get to California someday, but that day was very far off, and it would require outmaneuvering the all-powerful associates of the Southern Pacific to achieve it.

This was the state of play in February 1880: The Rio Grande was poised for far greater things, the Santa Fe left to ignominy, and Gould free to carry on just like always.

CHAPTER 17

A Whiskey Salute

BY THE TERMS OF GOULD'S "TREATY OF BOSTON," STRONG returned the Rio Grande to General Palmer at the stroke of midnight on April 4, 1880. Expecting to hold the Rio Grande for thirty years, Strong ended up running it for only seventeen months. First thing the next morning, the General's men resumed laying track from the Dead-line up to Leadville at a furious pace, keen for the silver revenues. Incredibly, the line was completed in just three months. On July 22, 1880, the first Rio Grande train rolled into town at last.

Once early passenger was a travel writer who declared that there was nothing like riding a railroad up into the Rockies. The traveler "has gone past it, gone over it, it may be; now he is going *through* it," he wrote of the experience. "The roar of the yeasty, nebulous-green river at his side, mingled with the crashing echoes of the train, reverberating heavenward through rocks that ride perpendicularly to unmeasured heights. The ear is stunned, and the mind refuses to sanction what the senses report to it." Clearly, the writer had no idea that the struggle to create that route had possibly been even more dramatic; if he'd looked out the window, though, he'd have seen the forts high up on either cliff side as evidence of the conflict.

The first train was pulled by a locomotive dubbed Fort de Remer. Draped with the requisite flags, it pulled two cars stuffed with dignitaries, headed up by former US president Grant and his wife. Unfortunately, the maiden voyage did not deliver any views at all, for the gorge was engulfed in dark clouds that unleashed a ferocious rainstorm along the whole route that delayed the train's arrival by three hours. By then, the vast, boisterous crowd of thirty thousand that had come out to greet it had long since turned into a drunken horde; the many tipsy celebrants saluted the incoming train with empty whiskey bottles when it arrived. General Palmer was not among them, leaving to Lieutenant Governor Haw Tabor the task of welcoming his former Union commander. Attended by thirty Civil War veterans, the volunteer fire department and five marching bands, Tabor stood amid an array of mounted police and military companies that sent off a hundred-gun volley of greeting.

A handsome barouche pulled by four black horses conveyed the Grants to the Clarendon Hotel for a formal banquet. Tabor sat beside the former president on the dais—much to her chagrin, Baby Doe had to sit among the other guests since Tabor's divorce from Augusta had not yet come through. A determined legal adversary, Augusta ultimately decamped to California, where she invested a good piece of Tabor's silver fortune in the Singer Sewing Machine Company, which proved to be of more lasting value.

The next day, Tabor took Grant down into the mines as he had Oscar Wilde to show off the various smelters and the spots where early gold was found. Afterward, Tabor sought a quiet tête-à-tête with his esteemed guest in his Opera House, but Grant begged off and left Leadville at daybreak.

In early August, Queen became the first Palmer to take a Rio Grande train up to Leadville on one of her increasingly rare visits to Colorado Springs. It made for an especially daring trip, since Queen was now, at long last, pregnant again, and six months along. Since the General was away yet again on railroad business, Queen brought along Alma Strettell, a friend from England. Concerned about Leadville's reputation for lawlessness, Queen had taken the precaution of tucking one of her husband's pistols into her picnic basket for the journey.

Both women were particularly keen to see the Tabor Opera House, then the most opulent building in Colorado. Opera had become quite the rage in the state as its fortunes increased, with over a hundred opera houses in operation. Queen and Strettell were dismayed to find the Tabor stood among saloons and bordellos.

Otherwise the trip was uneventful—until, as the two women returned by train to Cañon City, Queen felt a stabbing pain in her heart. Strettell managed to get her friend back to Glen Eyrie, where doctors determined that she'd suffered a heart attack. It was not fatal, but it was debilitating, requiring lengthy bed rest. Queen attributed her ill health to the high altitude and dry air, the very qualities that her husband believed to be restorative. It seems symbolic that, of all her organs, Queen's heart was the one most afflicted—and that it had failed aboard one of her husband's railroads, which had always been her rivals for his affection, traveling the route that he had almost literally killed for.

In her weakened state, Queen stayed on at Glen Eyrie long enough to deliver the Palmer's second child, a daughter named Dorothy. But

Dorothy Palmer as a toddler.

that was the end of Colorado Springs for her. After she left with Dorothy and Elsie, she never came back for any length of time. She was thirty.

▸·❖·◦·❖·◂

As for Strong, he refused to see the outcome of his battle for Leadville as a defeat. His Santa Fe would just have to grow in the Southwest. Even in the midst of the war with the Rio Grande, he'd been driving determinedly over the Raton Pass and into New Mexico. Early in January of 1879, a full year before the Treaty of Boston, he'd reached Las Vegas, New Mexico, a hundred miles inside the state. By then, Morley had surveyed much of the terrain, and decided it didn't make sense to go straight to the city of Santa Fe after all. The city did not have enough people or commerce to justify the high cost of building through the jagged Sangre de Cristo Mountains to get there. So, after the railroad crossed over Raton, he recommended skirting the mountains and continuing straight south to slip through the Glorieta Pass and then curve west to tiny Lamy on the flatland eighteen miles south of the city. To reach Santa Fe itself, Morley advocated pushing up a short spur up from there while the main line continued on to Albuquerque, bound for California.

Strong agreed, and Morley saw to it the work was done quickly. The first Santa Fe train climbed up to the city of Santa Fe from Lamy on February 9, 1880, exactly a week after the Treaty of Boston. Since no roundhouse had been built, the train could not turn around, and had to back into Santa Fe caboose-first, pushed, not pulled, by two locomotives because of the steep grade into town.

The train bore the usual load of dignitaries, most of them from the Santa Fe railroad. Strong was not among them, a clear sign that he viewed Santa Fe as just a way station, and hardly his ultimate final destination. The governor of New Mexico, Lew Wallace—a versatile character also known as the author of *Ben-Hur*—drove the track's final spike, and local citizens broke into a joyful fiesta in celebration.

It was quite a moment for what had once been a small Kansas line intended only to go from Topeka to Atchison. Under Strong, the railroad had realized that wild dream that Cyrus K. Holliday had shouted to the sky over a decade before. By now, Holliday had long since moved on from the railroad he founded, so he did not witness the train pulling in, either.

Santa Fe's population soon rebounded with a train, but Strong directed his attention to the railroad's interests farther down the line. Let General Palmer haul lead carbonate out of Leadville to his heart's content. Its silver would someday be exhausted, just as its gold had been. Strong would be selling something in his railroad towns that was inexhaustible, and only increase in value as the towns grew in number. And this something would not just transform the West. It would change all of America.

CHAPTER 18

Harvey Houses

THE SON OF A DESTITUTE TAILOR IN LONDON'S SOHO, Fred Harvey had come to New York City at the age of seventeen, in the spring of 1853. There he started work as a "pot-walloper," or dishwasher, at Smith & McNell's, a fine-dining establishment that had been started as a cake shop. Harvey went on to be a waiter at other high-end New York eateries before he left to take over a restaurant in Missouri. Soon after, he got in with the railroads, selling tickets, then shipping freight. Eventually, he returned to his first love, food, and never left.

Drawing on his experience with New York's haute cuisine, Harvey found a way to offer railroad passengers something else besides a destination and nice views en route—fine food, well served. By then, the Pullman Car Company had introduced dining cars that provided pre-cooked meals from Delmonico's, but the food was certainly not fresh, and the trains had no between-car "vestibules" that allowed passengers to come and go. Diners had to stay in the dining car the whole ride between stations if they wanted to eat on the train or choke down whatever gruel they could find at the depot once the train stopped.

Harvey had the idea of putting a proper dining room in the station. Passengers might have to come well before their train departed, or linger after it arrived, but either way, they'd be rewarded with an enjoyable meal to consume at their leisure in a nice place. Harvey tried to persuade the Kansas Pacific to test the idea at its stop in Denver, but

Fred Harvey.

they remained unconvinced. When he tried the Chicago, Burlington, a representative said he couldn't quite see it, either, but maybe Harvey should ask the Santa Fe, in 1875 a scrappy line trying to get deeper into Colorado. "They'll try *anything*," the rep told him.

Harvey had known the Santa Fe manager in charge of the Topeka station from his railroading days, and, as a favor, he let Harvey try an "eating house" there. It proved a wonder. A Topeka newspaper extolled it as "the neatest, cleanest dining hall in the State," and marveled at "crockery, cutlery and silverware" never before seen west of the Mississippi. Altogether, the paper said, it was "a luxury to set down to such a table."

This was promising, but Harvey's venture languished—until Strong came along in late 1877. Even as he was fighting the General for the Raton Pass and the Royal Gorge, Strong relished the idea of delivering a memorable dining experience to Santa Fe passengers. It seemed to him that it might somehow buff the image of his railroad. But the war with the Rio Grande had left him short on funds, so Strong told Harvey that if he could come up with the money, Strong would give him a prime location down the tracks at Florence, Kansas, and a goodly share of any profits. Since he'd put the financing on Harvey, Strong felt obliged to cede him creative control, and Harvey made good use of it. He didn't just start a restaurant; he built a lovely, stylish, memorable hotel to house it. There, he pulled out all the stops, luring William H. Phillips, probably the most famous chef in the country, to design the dining experience. Phillips had been top chef at the Globe Hotel by Philadelphia's Fairmount Park, serving a half-million meals to tourists coming in for the city's spectacular centennial exhibition of 1876.

Harvey offered Phillips the chance to create the restaurant of his dreams, and Phillips took it. He hired local fishermen, hunters, and farmers to provide a spectacular variety of fresh food straight to his kitchen. He served the meals off Sheffield china, with London stemware and napkins and tablecloths of Belfast linen, creating a dining sensation that was only a little less transporting than a Santa Fe train journey itself. A correspondent for a London sporting newspaper, drawn to Phillips's restaurant by reports of a hundred-pound catfish on the menu, simply could not get over the "marvels of luxury and neatness . . . suited to the most exigent or epicurean taste." Now wherever in America he was, he'd wonder if it was possible "to reach Florence and our compatriot Phillips in time for breakfast."

Phillips's restaurant was such a hit that Strong asked Harvey to put in another one just like it in Lakin, at the far end of Kansas. When that clicked, too, Strong had Harvey add another in La Junta, where the new line split off for Trinidad. Once the Santa Fe got over the Raton Pass, Strong allowed Harvey to put in his houses pretty much wherever the railroad went.

By now, Strong was so personally enthralled with Harvey that he started to invest in him personally. He bought ten thousand cattle in Harvey's private XY herd. He fully believed that Harvey was changing the meaning of train travel. It was no longer merely a matter of delivering passengers to a destination. Now it was about offering a memorable journey along the way.

And nothing would be more memorable than the first Harvey House to go up in New Mexico. Harvey named it the Montezuma after the Aztec king, a nod to the scale of its aspirations, and placed in the territory's newest attraction, Las Vegas. Located by a mineral spring where the outlaws Jesse James and Billy the Kid sometimes went for a soak, it became a high-end resort for the health conscious.

Built on the scale of a full town, the Montezuma rose off the prairie like a mirage. Done in the turreted Queen Anne style, it soared four stories high, and covered nearly a hundred thousand square feet to make it the largest wood-frame building in the United States. It featured 270 guest rooms, a dining room for five hundred, a casino, billiard hall, and six varieties of medicinal baths. The liveried staff was drawn from eastern hotels that were steeped in the European traditions of proper service, and, following Phillips's principles, the restaurant featured a range of food made from fresh ingredients. The Montezuma's menu included green turtle soup, broiled teal duck, Muscatine ice, and even "sea celery harvested by pearl-diving Yaqui tribesmen," all cooked with a flair.

Although located in the desert of New Mexico, the grounds featured an English-style park with shade trees and colorful flowers, all arrayed around a lavish fountain, plus lush lawns for tennis and croquet, an archery range, and a menagerie, all of it lit up through the night by thousands of gaslights.

In short order, the Montezuma drew its guests from every corner of the United States, and from many of the countries around the world, all of them arriving at the hotel's front door, courtesy of the Santa Fe railroad.

This introduced something new to the West—"excursionists." These were not the temporary residents the General brought to Colorado Springs, but people coming for just a brief stay of a few days before pushing on. They weren't travelers, but consumers of travel, a commodity that replaced hardship with comfort, and necessity with luxury. The transformation made the West seem a lot less forbidding.

Eventually these excursionists would be called tourists. To serve them, Harvey brought in women—a particular kind of women, actually. The enticing, young, unmarried variety. As the saying went, there were "no *ladies* west of Dodge City and no women west of Albuquerque." Well, no more. If the raw masculinity of the West gave it so much of its rough cast, the "Harvey Girls," as they were termed, would change all that.

Harvey turned to them when he found that his diners were put off by the black male wait staff he'd been routinely employing. Subject to abusive vulgarities from the Confederate-sympathizing cowboys who were the Harvey Houses' principal western clientele, the black waiters had taken to carrying six-guns to protect themselves, and such weaponry conflicted with the genteel tone Harvey was trying to set. The whole thing blew up for good in 1883, when a brawl broke out at the Harvey House in Raton. Apparently, some black waiters had taken too many insults, and pulled their guns. Some knives came out and, as the manager callously reported to Harvey, "several darkies had been carved beyond all usefulness."

Harvey realized that he could avoid such problems by importing waitresses from civilized places like Topeka or Kansas City. At the very least, they weren't likely to "get likkered up and go on tears." He turned his Girls into models of cleanliness, both in personal hygiene and in morals, as he insisted they wear spotless white and black uniforms, and no makeup, lest they be mistaken for those *other* ladies of the night— although he made sure those uniforms were formfitting. And, being unmarried, they carried a hint of availability.

Tragically, the Montezuma burned to the ground in an electrical

Some Harvey Girls from the 1920s.

fire before the end of its first year. Strong rebuilt it, only to see it burn down once more. When he rebuilt it yet again, it remained standing, but it was never a success, probably because it wasn't a true Harvey creation, which was never fancy, just good—and better than anything else around.

Harvey continued to put up eateries, many of them with superior hotels attached, throughout New Mexico, in places as far flung as Deming to the south, Raton to the northeast, and Gallina to the west. Soon, there were ten Harvey Houses altogether. Eventually, there would be a Harvey House every hundred miles along the Santa Fe tracks, eighty-four altogether. With his restaurants, Harvey brought standards to the West. Harvey Girls knew to slice the bread *exactly 3/8ths of an inch thick,* to squeeze the orange juice *only after it was requested,* to indicate to the "drinks girls" that the customer wanted coffee *by leaving the cup in the saucer,* but if the customer wanted ice tea, *to invert the cup and leave it leaned up against the saucer's edge.* When Harvey inspected each restaurant, he took out his handkerchief to swipe the tops of doors and windows to check for dust. If his standards weren't met, he could be savage. Any chipped plates he'd smash on the floor. And any mislaid tables he'd tip over.

It was a natural outgrowth of industrialization, this notion of perfect uniformity. Harvey Houses made for the nation's first chain restaurants, and as they spread across the West, they brought unity to a sprawling outback, and a welcome bit of gentility, too.

⊳—⊶—○—⊰—⊲

While both Strong and the General sought a certain elevation in the travel experience, only Palmer associated it with exclusivity. Strong was not trying to appeal to a privileged few, but to a receptive many. His impulse was democratic, a matter of numbers. Strong always trusted volume.

The Santa Fe was not the first railroad to carry tourists, but it was the first to cater to them. The Harvey Houses were the first to develop the postcard for their guests to show off the local scenery to friends back home. Harvey soon added full tourist books that gave the West a romantic gloss for eastern consumption, and organized tours of the nearby countryside playing up the local color.

To enhance a sense of place, he displayed the indigenous architectural styles of the Southwest in his hotels, rather than adopt European standards as Palmer had done. In the city of Santa Fe, for instance, Harvey built La Fonda in the Spanish pueblo tradition, solidifying the adobe character of the city. And he made Native American culture a selling point. At some of his hotels, Harvey organized "Indian Tours" of the nearby Indian lands, where he arranged for natives to be on display, and created in-hotel retail shops to sell the jewelry, artwork, and other artisanal creations of the local tribes. He used an Indian thunderbird emblem for the Harvey House logo, and slapped it on every plate, bowl, and piece of cutlery in his eateries. He also brought in anthropologists to record the traditional ways of these vanishing tribes and encouraged artists and photographers to capture their spirit before it was lost. The movement ultimately brought artists such as Georgia O'Keefe to Taos.

As Strong pushed ever deeper into the West, he gained for his railroad the Harvey House aura of service—reliability and good taste. Advertising "Fred Harvey Meals All the Way," the Santa Fe made clear it was not just another railroad. And Strong was now poised to take the Santa Fe brand all the way to the sea.

Railroad passengers arriving in Los Angeles.

PART THREE

LOS ANGELES

CHAPTER 19

The Pueblo

EVEN THOUGH PALMER AND STRONG PUSHED ON IN SEP-
arate directions, they were not finished with each other since they were
both in pursuit of the same ultimate objective—to reach the Pacific.
Much as each man strove to create places of value along his route, be
they at Colorado Springs or Albuquerque, they both recognized that,
ultimately, a route is defined by its destination. The first transcontinental, that Pacific Railway, was conceived on this principle, and Strong
and the General's lines were, too.

While both men had flirted with the notion of hitting the Pacific at
the Mexican shore, they both took California as the western ultimate.
It was for all Americans the place of the future, the final frontier, the
sunny paradise by the sea that represented every pioneer's ultimate
objective.

If all roads led to Rome in the days of the Roman Empire, all railroads led to San Francisco in the California of 1880, and quite sumptuously. To arrive in style, the Central Pacific offered a Pullman Palace
Car dubbed the Russian Grand Duke, "perhaps the most commodious
and perfect manner in which any one has ever traveled by rail," as one
promoter plumped it. By then, California was "a country by itself," a
British traveler observed. "And San Francisco a capital." Thanks to the
Central Pacific, the city was so infused with money, style, population,

and influence that the rest of the state was obliged to bow to it—and run its Central Pacific lines there. As it grew, the CP became better known as the Southern Pacific, a corporate monolith that was a force to be reckoned with everywhere in the state. The Southern Pacific had sent a line down from San Francisco through Bakersfield to Los Angeles in 1876, but only after charging the city an extortionate cash subsidy of five percent of the total value of all its real estate.

The L.A. venture was part of a grand strategy by which the Southern Pacific monopolized all of California, a feat that no other railroad in any other state had come anywhere close to achieving. And California, of course, was the largest state of them all. By 1880, the Southern Pacific either had built enough of its own lines, or bought enough of its competitors, to control eighty-five percent of the tracks in the state, a whopping 2,340 miles altogether. Its stock was worth $225 million, over two-thirds of the federal budget. To Strong and the General, that presented a brutal fact: unlike the rest of the Southwest, California already had its network of trains, and it was a network designed to keep any other railroad out.

As a monopoly, however, the train ran as its owners dictated, not as its potential users would have liked, and exploited their power to set rates to the company's benefit, without any acknowledgment of a market rate, for there was no market. There was only the Southern Pacific. So it determined by corporate fiat where goods and passengers flowed. If San Francisco was the almighty Rome of this railroad empire, and Los Angeles a pathetic backwater, it was because the lords of the Southern Pacific wished it so. L.A. was never to challenge the capital's preeminence, ever.

When the General had created his Rio Grande in Colorado a decade before, the state was all his for the taking. And there was relatively little to Colorado, with just one city of any size, and the state government barely a year old. California, by contrast, had been a state since 1850, and its leading city of San Francisco had more than 230,000 people, making it the ninth largest metropolis in the country.

Poised on steep, fog-shrouded hills overlooking the gentle waters of its namesake bay, San Francisco had enough swanky hotels and bus-tling nightlife to give it a good deal of the panache of New York and Chicago, with a fair amount of the exotic thanks to its proximity to Asia. One writer called it a combination of Peking and Marseilles. To Oscar Wilde, it was "the Occidental uttermost of American civiliza-tion." When he came into California from Leadville, he had stopped only in San Francisco, and by one account dressed for the occasion in "a Spanish sombrero, velvet suit, puce cravat, yellow gloves, and buckled shoes." He made straight for the Bohemian Club, a sparkling literary and artistic enclave infinitely more fascinating and outré than anything else west of New York City.

San Francisco, of course, had been created by the epic gold rush of 1849. (By 1900 it had pulled out of California a staggering $30 bil-lion worth in today's money.) The first strike had occurred just a few months before Mexico ceded to the United States half its territory after its bitter defeat in the Mexican-American War. Technically, that treaty created the state of California, but in every other way, it was made by gold. California was branded the Golden State, a place of instant good fortune, and as such a beacon to anyone who aspired to get rich quick— and that appeared to be pretty much everyone in the country. America had always offered golden opportunity. California delivered on it.

The first rush began by accident when a carpenter was digging out a foundation for a sawmill outside Sacramento and noticed something glittery. He managed to keep the news quiet for over a year, until May 1849, when a local paper picked up on it and created a national stam-pede. "What crowds are rushing out here for gold," wrote one aston-ished California minister. "What multitudes are leaving their distant homes for this glittering treasure!" Most prospectors came by ship all the way around Cape Horn, or to the Isthmus of Panama, crossing it by wagon and then boarding another steamer to San Francisco from there. Only a few braved the overland route over the mountains and across burning desert. However they traveled, all endured long, peril-ous journeys that revealed a lot about the intensity of the prospectors' desire for a fortune.

The numbers were staggering. In 1849, there'd been only fifteen thousand people in all of California, top to bottom. By 1853, there were more than three hundred thousand. Most of them settled in the northern gold-bearing towns of the Sierra Nevada mountains, but San Francisco provided the harbor for all who came by boat; with all the port traffic it rose in just one year from a lonely outpost of one thousand to a thriving city of twenty-five thousand. It was also the only decent place for miners to settle after they'd made their poke.

The gold fever was the quintessence of the hypercapitalist age—which was, as Twain observed, gilded. If corporate America divided the country into those who worked and those who profited off others' work, this provided at least the possibility of a third category—those who profited wildly off their own efforts. But the profit was still the defining quality. One might be able to discern softer sentiments behind the industrialization that led to the transcontinentals—a desire for the new, a pleasure in ingenuity, a delight in the western landscape—but by far the most dominant and the most universal emotion lay in the personal quality that Adam Smith delicately termed "self-interest," otherwise known as greed. It was greed and nothing but greed that brought the prospectors thundering into California, and there had to have been a lot of it to bring so many people so far. Later, it brought the train men, too.

Since the California gold was confined to the northern Sierra Nevada, it left the south relatively untouched. The Mexican influence had never much extended into northern California, so far away from Mexico City, and was crushed when the gold-seeking Argonauts piled in, drawing in the cultures of their faraway origins with them. Mexico remained more evident farther south, if only because so few settlers had come in to replace it.

Although southern California had been claimed for Spain by conquistadors in 1542, it had not developed much since. By the time Strong and the General started to focus on California in 1880, Los Angeles was still a dusty Spanish pueblo of eleven thousand people by

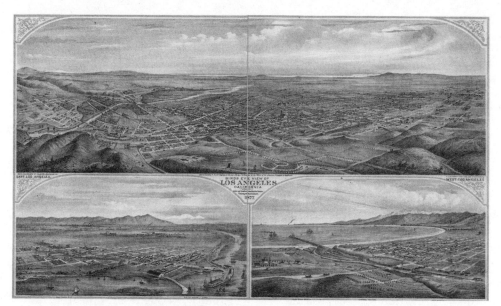

Views of Los Angeles in 1877.

the sea. It featured a smattering of forgettable adobe buildings along unpaved streets, the irregular blocks giving way to orange groves that rose up helter-skelter into the surrounding hills. The Southern Pacific line into the city was so little used that the "prevailing temper" of this sleepy, seaside place, one observer wrote, was "the soft and comfortable spirit of *mañana*."

The surrounding countryside offered little more. For decades, bee ranching had been a major industry. A typical homesteader could sell forty thousand pounds of honey a year— "whiffs of honey-laden air came from stretches of chaparral thick with wild bees," declared a visitor—and the more ambitious agriculturalists grew odd-lot products like walnuts, sheep, and silkworms. Northern refugees from the East Coast felt drawn to the abundant sunshine to create farming colonies wherever they could bring in enough irrigation. One of the more notable was started by emigrant Mormons in San Bernardino, sixty miles east of Los Angeles. Other community growers had started other fledgling towns like Riverside and Long Beach.

But easily the most impressive was Pasadena, which bordered Los

Angeles to the northeast. It was founded in 1874 by a small group of midwesterners who'd had one too many brutal winters and formed the San Gabriel Orange Grove Association. Like Palmer in Colorado Springs, they laid out a town of smartly parallel streets intersected by wide avenues, with handsome houses on plots of perfect squares. Instead of the Rockies, however, Pasadena offered a view of sunlit orange and lemon groves. Soon the town was filled with improving institutions, including a one-room schoolhouse that added a meeting hall on a second floor and became a small college, which ultimately turned into the California Institute of Technology. There was also the Pasadena Library, the Village Improvement Society, and a Literary Society. With its dry, healthful air, Pasadena drew wealthy consumptives much as Colorado Springs had. Initially, the sick visitors put the town in a bad light when spotted, as one observer had it, "coughing upon steps or leaning against walls with eager, anxious faces," but ultimately, a retreat called the Sierra Madre Villa offered them a comfortable retreat. At first it was just a large, private house on five hundred acres, planted with orchards and ornamental trees. When it proved popular, a spacious verandah was added to turn it into the town's first proper hotel, offering horseback riding and dances for the guests, who were fortified by the cheerful climate. Initially, they just stayed a week, but before long guests were returning annually for the full winter season. The impressive numbers won the notice of a Boston travel agency that saw the beginnings of a tourist destination for easterners fleeing winter. Still, while the town attracted some illustrious high-enders like the ubiquitous Ulysses S. Grant and the naturalist John Muir, it did not quite pop. In 1880, the downtown was still a matter of a few drab wooden-frame storefronts facing a forlorn street.

To grow into anything more, Pasadena needed a train.

CHAPTER 20

The Big Four

WHILE JUST ABOUT EVERY OTHER RAILROAD IN AMERICA was customarily topped by just one man, be it Strong, Palmer, or Jay Gould, the Southern Pacific had four, the "Big Four" as they were known, when they weren't more dryly referred to as "The Associates."

The Big Four had all once been Sacramento shopkeepers who'd come west in the gold rush, only to realize that the real money was not likely to come from panning for gold, but in selling dry goods to the fools who didn't know any better. "I never had any idea or notion of scrambling in the dirt," said the best-known of them, Collis P. Huntington. When the gold showed signs of petering out, the Big Four turned to the next big thing: the Central Pacific Railroad, which, unlike the gold, could be all theirs, every bit of it.

With the Federal government on the hook for so much of the construction money, the Big Four needed to scrounge up just $300,000 among them to buy a controlling interest in the railroad and win a broad swath of federal land on either side of the tracks. That land amounted to one-eighth of the state—the most valuable one-eighth, since it was the portion served by the railroad. Once they snapped up the subsidiary lines to control the state's traffic, they effectively took charge of the state itself. Even in its earliest incarnation as the Central Pacific, the company was called the "third party" that actually ran the state, topping whichever of the two political parties foolishly imagined it was in power. It was said that before an elected California official went to Washington, the Central Pacific placed a collar around his neck bearing the words "Central Pacific" "so if he is lost or strayed he may be recaptured and returned to his lawful owners." When the state created a three-man railroad commission to investigate the monopoly prices imposed by the Big Four, two of them were on the Central Pacific payroll. Rates, needless to say, remained untouched.

On the all-important greed scale, Mark Hopkins ranked lowest of the Big Four. He was a gaunt, lisping vegetarian of abstemious habits and a bookkeeper's caution. He was also the first to go, dying in his sleep in his private railroad car in 1878. Then came Charles Crocker—or Charley, the only Associate personable enough to get a nickname—a former newsboy who turned lazy with wealth. "His feet are more often on the desk than under it," the *San Francisco Examiner* once wrote. Shortly after, Crocker cashed out and went off on a two-year sojourn to the honey spots of Europe before buying back in. He was best known for putting up a $2.3 million house on a solid block of San Francisco's Nob Hill, where he installed a forty-foot "spite fence" facing his neighbor, a Chinese undertaker who'd refused to sell him his parcel. (The undertaker retaliated by placing a coffin atop his roof and flying over it a flag of a skull and crossbones.) Next came the handsome, confidently full-bearded Leland Stanford. He had a touch of public-spiritedness, trying for the governorship before becoming a US senator, as well as enlisting the early photographer Eadweard Muybridge to take the now-famous shots of galloping racehorses that led to moving pictures. He also created Stanford University to memorialize a son who died young. He had a gargantuan Nob Hill mansion of his own, albeit a more tasteful one, with Italianate architecture and a stone entrance hall inlaid with signs of the zodiac in black marble.

Collis P. Huntington was without question the greed champion. The Great Persuader to some, the great conniver to others, he stood a robust six feet, with metal-gray eyes, and dressed in funereal black, as if preparing to bury his many enemies. The only speck of cheer on him was a gold pinky ring. If there was ever a trace of human sympathy on his face, his heavy beard concealed it. Born to a broken-down farmer

Collis P. Huntington.

in Poverty Hollow, Connecticut, Huntington went West via Panama
to get in on the gold rush. But there were no carriages waiting to carry
the ship's passengers across the isthmus, and, stranded in the boiling
heat for two months, passengers fell to famine and disease until Huntington hacked thirty-nine miles through the jungle to find food to sell
to starving customers for a three-fold markup. The money bankrolled
his first store.

With Huntington leading the way, the Big Four used their railroad
monopoly to preserve their influence, forcing communities to pay exor-
bitant fees for tracks, and then charging outrageous prices to use them.
And death to any invader. The first to try was Tom Scott, the domi-
neering head of the Pennsylvania Railroad, then the country's largest
train company. Two years before the Pacific Railroad was complete,
Scott wanted to join the Pennsylvania to the Central Pacific at Denver to create the nation's second transcontinental. Huntington got his
friends in Congress to kill his bid.

In 1876, Scott was back at it. He started the Texas and Pacific with the idea of sweeping west across Texas and Arizona to hit San Diego along the 32nd parallel.* Because of the vast expense involved in building such a route over hot, barren desert, Scott figured he'd need a federal land grant to pay for it—but Huntington had extracted one to build a line of his own across the state from Los Angeles, and he didn't want a competitor. Huntington hired a political fixer, David D. Colton, to shut down Scott's T & P in Congress, and started building east from L.A. in 1877, daring Scott to double the SP's tracks and incur the full expense of construction for half the potential revenues. Before Scott could lay down a rail, Huntington had reached the Colorado River at Yuma, Arizona, in 1880, near the Mexican border. Needless to say, he never intended for this route to turn L.A. into a more populous or prosperous city. He planned for it to lead to San Francisco, the capital of his railroad empire.

At Yuma, Huntington encountered a rare confounding obstacle. It turned out it was mostly *Fort* Yuma, a military base under the direction of the secretary of war George McCray, who denied Huntington passage. Unfortunately for McCray, all but three of the fort's soldiers were off fighting the Nez Percé Indians, so Huntington did not stop. He threw a bridge over the Colorado and ran tracks right through the fort. Learning about this, McCray was apoplectic. After he reported the events to Washington, Huntington felt obliged to meet with President Hayes to explain his side of things. "He was a little cross at first, said we defied the government, etc.," Huntington blithely reported. "I soon got him out of that belief." In fact, he got Hayes to see the humor of the situation and soon had him roaring with laughter. Hayes told Huntington to keep on building. "That will suit me just fine," he said.

---

* The future Confederate president Jefferson Davis had initially championed the 32nd parallel when the Congress was in search of the best route for the Pacific Railway and he was a Mississippi senator. He argued, correctly, that such a southern route would avoid the paralyzing snows and mountains of a northern one. It would also have yielded considerable benefits to his treasured South, which doomed its prospects as the Civil War approached.

Scott never did get his land grant. His T & P remained struck in Texas while Huntington secured what he perceived to be his due. To lay claim to the federal land he'd been awarded in south-central California, he dispatched his minions to take it by force from the ranchers who lived there, even though they'd worked the land so long they had reason to think of it as theirs. When the ranchers got no relief from the courts, they started burning down the ranches seized by the SP. Imagining that Leland Stanford had a common touch, Huntington sent him out to try to calm things down. It didn't help that Stanford arrived by a sumptuously furnished Southern Pacific private railroad car. A United States marshal named Poole was dispatched next to work something out. Poole brought an SP man with him, along with a pair of real estate speculators who were prepared to pay up to forty dollars an acre for land that had gone for $2.50 before the Southern Pacific arrived.

Fifty ranchers thundered up to confront the group. When the marshal climbed down off his buggy to have a talk, the men asked him to hand over his pistol. Poole said he'd rather not. The ranchers pointed several gun barrels at his head. There were words. A skittish horse knocked into Poole. A gun went off, and bullets flew. By the time it was over, eight men lay dead. Miraculously, Pool and the SP man emerged unscathed.

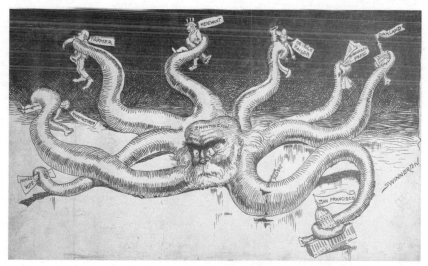

Huntington's Southern Pacific as an octopus.

Rather than express remorse, Huntington secured convictions against six of the surviving ranchers for obstructing justice. This blackened the image of the Southern Pacific forever. It was why Frank Norris, in his epic novel about the Southern Pacific, *The Octopus: A Story of California*, famously called the railroad a "galloping monster" and "terror of steel and steam, with its single eye, cyclopean, red, shooting from horizon to horizon."

This was the railroad that any other railroad was up against if it sought to break into California.

CHAPTER 21

"A Terrible, Single-handed Talker"

BY THE TIME OF THE SAN JOAQUIN VALLEY SLAUGHTER IN 1880, the Santa Fe already had a presence in the state. Incredibly, even as Strong was building over the Raton Pass and fighting his way up the Royal Gorge, he was also scheming to take California from the inside, and then to work his way east to be joined by the line he was projecting from Raton into New Mexico.

Even in a country as big as the United States, Strong understood that growth had to stop at some point. Nothing goes forever, not even a railroad, but Strong was determined that the Santa Fe would reach the far end of the country before it stopped.

In 1880, it looked like the end might be in San Diego. This was quite by chance. San Diego was desperate for a train, but Huntington had no interest in delivering one, since it would only build up the San Diego port into a threat to San Francisco's. When the Southern Pacific finally did build a line down through the Central Valley to Los Angeles, it stopped there, leaving San Diego in the position of Trinidad back in Colorado, tantalized by a train that was being withheld from it.

The citizens of San Diego were furious. If the Southern Pacific would not deliver a train, they'd find a company that would. For the purpose, they turned to a slight, scraggly bearded former carpenter named Frank A. Kimball. He'd come from New Hampshire to start a town he called

National City, just south of San Diego, that was basically all his ranch. He built a post office, schoolhouse, and wharf, and he put up the first house with a bathtub. To get a train, Kimball had persuaded Tom Scott to bring in his ill-fated Texas and Pacific. After Huntington blocked the deal, Kimball tried Jay Gould. He replied with his famous line, "I don't build railroads. I buy them." Finally, in 1879, Kimball turned to the Santa Fe and journeyed to Boston to court the company in person.

Kimball could be "a terrible single-handed talker," a friend of his noted, and he pestered Nickerson and the Santa Fe board for a full three months before Nickerson finally said yes, perhaps just to be done with him. Still, he had offered the Santa Fe a very sweet deal: seventeen thousand undeveloped acres of National City, plus two full miles of harbor frontage, and almost five hundred lots, seemingly enough to guarantee the train's success whatever happened. It was all in the service of something that to Strong was infinitely more valuable—entrée into California across the bottom of the state, well away from the clutches of the Southern Pacific. From his early experiences in New Mexico, Strong was fully aware of the lengths that the Southern Pacific would go to protect its California exclusive. Strong hoped that by starting in at San Diego so far to the south, he could get the train going before Huntington found out about it. To conceal the Santa Fe's involvement, Strong would incorporate the line as the California Southern to make it look like a local affair the way he had deceived Palmer with the Cañon City and San Juan line while he prepared to take the Royal Gorge. The new company was incorporated on October 12, 1880, just eight months after the Treaty of Boston forced Strong to look to the Southwest for the future of his railroad.

The idea was to run a line up from National City to San Diego and then up the coast, before veering inland to San Bernardino and running east from there. It proved no easy job, though. All the rails and heavy equipment for the new line had to be shipped from Europe, around Cape Horn. The first shipment was delivered by the British four-master *Trafalgar* in March 1881, and workmen methodically laid track up to Fallbrook Junction, fifty-five miles north, by the beginning of the new year. From there, the line angled up the Santa Margarita River through the treacherous Temecula Canyon. Locals warned that flash floods routinely crashed through, destroying everything in their path, but Strong needed to forge on before Huntington could stop him, and the California Southern was in too much of a rush to pay heed. Deploying a regiment of two thousand Chinese workmen, some of them veterans of the Central Pacific's drive east, the California Southern built up frantically through the canyon, and then rushed on to the town of Colton, fifty miles on, that August, with plans to go north through the San Bernardino Mountains from there.

By then, the Santa Fe's secret connection was out. Nickerson had left the Santa Fe to take over the presidency of the California Southern—leaving T. Jefferson Coolidge to replace him—and its board was thick with the Kidder, Peabody people who backed the Santa Fe. There was no doubt about the Santa Fe's involvement. Things came to a head at Colton, where the California Southern had to cross the Southern Pacific tracks that Huntington had put in that headed to Yuma on their way to El Paso, naming the town for his political fixer. The SP tracks at Colton would be his Dead-line in his fight with this California Southern, the end of its line.

Conventionally, to cross a preexisting set of tracks, an incoming train relies on what's called a "frog." It's a portable, prefabricated bridge that allows the new train to leap over the old one. Huntington would have none of it. He had a friendly sheriff seize the frog before it could be installed and then erect a fence along his tracks to keep the California Southern well clear. It was the Royal Gorge redux, and, as before, the action shifted to the courts. The California Southern appealed for a judge to uphold its right to cross the Southern California tracks, but with Huntington's iron hold on the state judiciary, it took a full year for a judge to deliver the decision. In a rare act of judicial courage, he gave the California Southern legal permission to proceed. But when the CS workmen arrived to install the frog, they found a massive Southern Pacific locomotive pulled up at the crossing point, blocking the work. By then, the citizens of Colton and San Bernardino were thoroughly

A well-decorated California Southern arriving in San Bernardino.

sick of the Southern Pacific. In a fury, they set upon the engineers with clubs, forcing them to move the locomotive out of the way.

Up went the frog, and over went the California Southern on fresh tracks all the way to San Bernardino. When the first train rolled in on November 16, 1883, its locomotive was gaily festooned with evergreens, flowers, and corn stalks, and its cowcatcher piled high with squash, all of it to represent the bounty the new train line would bring all along the line.

Then—disaster. Forty inches of rain fell on the Temecula Canyon, sending down a raging flood that ripped up eight full miles of the California Southern's track and sent them down the canyon and so far out to sea that bits of its bridges washed up along the coast eighty miles away. It was exactly what the locals had warned. By then, the new line was so far over budget, Nickerson couldn't afford to rebuild the lost track, let alone go any farther to get out of the state. By 1884, the California Southern had stalled out before it even reached the San Bernardino Mountains. It was a road to nowhere.

CHAPTER 22

Guaymas

DESPITE THESE OBSTACLES, STRONG WAS NOT ABOUT TO give up. To him, it was simple: The Santa Fe was going to California. Period. In fact, he'd already been working on a backup plan. In case he couldn't bring a train *out* from inside California, he'd send one *in* from outside California. All the same to him. Unflappable by nature, he was prepared for any eventuality.

He'd been developing this alternative approach long before he even got going with the California Southern. It was a scheme that had stemmed from his original ambition to glide west once he crossed over Raton into New Mexico. He had, after all, been continuing on that course, popping in Harvey Houses, from Lamy, where he delivered that spur to the city of Santa Fe; through Las Vegas and Albuquerque. By the end of 1880, he'd gotten all the way down to Rincon, due south of Albuquerque near the bottom of the state. Now he veered fifty miles southwest to a new town called Deming, where he hit the Southern Pacific's tracks, the ones that Huntington had set through Yuma, Arizona, to ward off Tom Scott. At this point, six months before the Southern California was temporarily stymied by the Southern Pacific's tracks at Colton, Strong thought there was a decent chance that Huntington would be reasonable and let him share the SP rails into California.

Indeed, Huntington wasn't particularly worried about the Santa Fe just then. A little line out of Kansas that had just lost the rights to Leadville—where was the threat? It was Huntington's fellow Associate

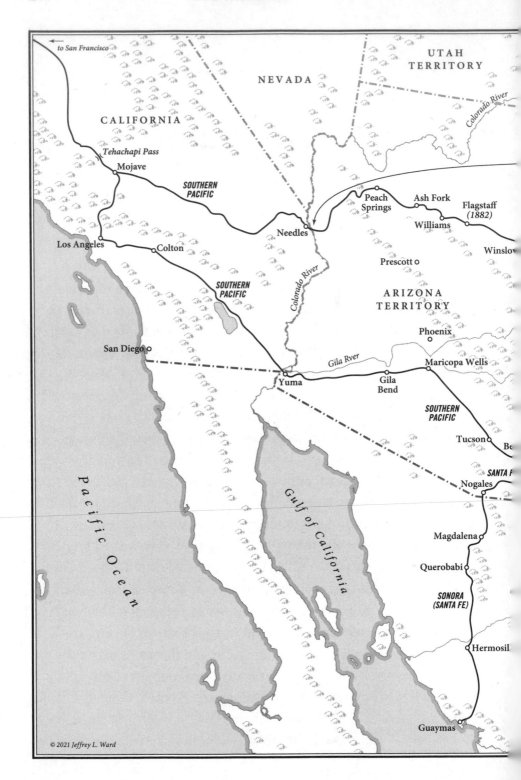

to San Francisco

UTAH
TERRITORY

NEVADA

Colorado River

CALIFORNIA

Tehachapi Pass
Mojave

SOUTHERN
PACIFIC

Peach
Springs
Ash Fork
Williams
Flagstaff
(1882)

Needles

Winslow

Los Angeles
Colton

Prescott

SOUTHERN
PACIFIC

Colorado River

ARIZONA
TERRITORY

San Diego

Phoenix

Gila Rver

Maricopa Wells

Yuma
Gila
Bend

SOUTHERN
PACIFIC

Pacific Ocean

Gulf of California

Tucson

Be

SANTA F
Nogales

Magdalena

Querobabi

SONORA
(SANTA FE)

Hermosil

Guaymas

© 2021 Jeffrey L. Ward

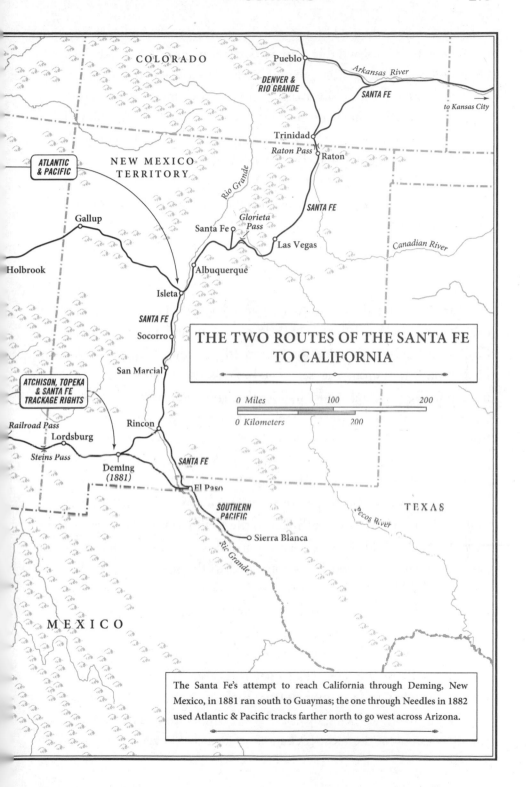

THE TWO ROUTES OF THE SANTA FE TO CALIFORNIA

COLORADO

Pueblo

DENVER & RIO GRANDE

Arkansas River

SANTA FE

to Kansas City

Trinidad

Raton Pass

Raton

ATLANTIC & PACIFIC

NEW MEXICO TERRITORY

Rio Grande

Gallup

Glorieta Pass

Santa Fe

SANTA FE

Las Vegas

Canadian River

Holbrook

Albuquerque

Isleta

SANTA FE

Socorro

San Marcial

0 Miles   100   200

0 Kilometers   200

ATCHISON, TOPEKA & SANTA FE TRACKAGE RIGHTS

Rincon

Railroad Pass

Lordsburg

Steins Pass

Deming (1881)

SANTA FE

El Paso

SOUTHERN PACIFIC

Pecos River

TEXAS

Sierra Blanca

Rio Grande

MEXICO

The Santa Fe's attempt to reach California through Deming, New Mexico, in 1881 ran south to Guaymas; the one through Needles in 1882 used Atlantic & Pacific tracks farther north to go west across Arizona.

Charley Crocker who told him to watch out, the line had a lot of Boston money behind it, and this man Strong was relentless. Nevertheless, Huntington decided to be lenient and let the Santa Fe join the Southern Pacific at Deming with a silver spike and a big ceremony on March 8, 1881. For a delicious moment, Strong thought he had indeed achieved the impossible—a route to the sea to make his Santa Fe the country's second transcontinental line. But then Huntington had a change of heart. Before the week was out, he imposed such a hefty surcharge on Santa Fe trains going to California on the Southern Pacific tracks that one railroad writer called it a "prohibitory tariff, rendering the way unuseable." The right to California was the Southern Pacific's alone.

⊳┄⊲┄○┄⊶┄⊲

But Strong wasn't done. He had *another* backup plan, a backup to the Deming backup. If he couldn't build out from California on the California Southern, or ride into it on the Southern Pacific tracks, he'd approach it from the sea. He'd break the Southern Pacific's monopoly by reaching a port in Mexico, and then running ships to California from there. And he'd rely on those Southern Pacific tracks from Deming to do it—not all the way to California since the surcharge made that prohibitive but 170 miles west from Deming to Benson, Arizona. From there, he'd cut south and build straight down and down to hit the Pacific on Mexico's western coast at the port of Guaymas on the Gulf of California. He could connect to the California coast by steamer from there. He could also reach Australia and the Far East, saving shippers well over a thousand miles compared to the sea route from San Francisco. He'd have a transcontinental all of his own, a real one, the western portion under his exclusive control, except for the stretch of shared SP track between Deming and Benson.

To Strong's surprise, Huntington posed no objection. He let Strong use the Southern Pacific tracks to get to Benson and told him if he wanted to build south to Guaymas from there, go right ahead. Huntington was, as ever, not being nice. He was being shrewd. The Mexican gambit was one of the most cockamamie ideas he'd ever heard. Why would an American company ever ship anything through a foreign country like Mexico? No matter how many miles it saved, it was not worth the hassle of dealing with the politics, bureaucracy, pesos, culture, and Spanish language of Mexico. Let Strong build his tracks down to this Guaymas—it would be his route to bankruptcy. Meanwhile, he would ride the Santa Fe's tracks east. The Southern Pacific wouldn't deliver Strong a transcontinental in any meaningful sense, but it would give Huntington one.

When General Palmer first read about Strong's plan in the newspapers at Glen Eyrie, he must have felt a slow, simmering rage rise within him. Strong was shamelessly pilfering the very idea upon which General Palmer had based his Rio Grande railroad. This was the *General's* plan! His Rio Grande was to be the north-south line, and Strong's Santa Fe was to be the east-west one. Now, Strong was going to be the north-south line after all? It was too much! Would Strong never stop stealing from him?

In fact, Strong had been thinking of the Guaymas option practically from the moment he took the job with the Santa Fe—which may have also been the moment that he learned of Palmer's interest in it. Back in 1878, he had sent the inestimable Ray Morley to plot a route to what Morley called a "sleepy fishing port" on the western coast of Mexico, two hundred sixty miles down from the American border at Nogales, Arizona. To get there, Strong created a new subsidiary, the Sonora Railway, named for the Mexican state it passed through and quietly secured a "cession"—or legal permission—from the Mexican president, Porfirio Díaz, to build into the country. Once a revolutionary, Díaz had swept into elective office in 1877 by offering Mexico a path to prosperity along the railroads. (After his four-year term was up, he took power for three decades more as a dictator.) Díaz offered to contribute toward construction costs much as the American government had initially for its first transcontinental, awarding the Santa Fe some public land, including any coal miles, salt deposits, and spring water that lay on it. The only catch was that the Santa Fe would have to prove the land really was public. It was far from a certainty, and the company was not likely to have an easy time persuading ranchers to sell them any of their private land for a reasonable price if they knew its value would jump with a train.

Overall, the logistics were a horror. Just as the California Southern had in San Diego, the Sonora Railway would have to bring in all the rails and equipment by ship around the Horn—and, off on the coast, far from the federal government in Mexico City, Sonora was rife with murder, ransom kidnapping, thievery, and rebellion that the state was nearly powerless to control. Few members of the Mexican labor force had worked with construction equipment beyond a shovel, and almost none had ever even seen a train, for there was only one in their entire country. It had been grandly called the Mexican Railway, though it only joined Mexico City to Veracruz on the Gulf of Mexico, and, ultimately funded by the British, had taken thirty-seven years to build. Not an inspiring precedent. The truth was, Mexico was frankly leery of an American train, and for good reason. In the Mexican-American war of 1846–48, an American regiment had fought its way into Mexico City, forcing Mexico's surrender. What if American troops could sneak swiftly into the country by train?

Much as Mexicans feared American military might, however, they envied American prosperity. The ambivalence showed when Porfirio Díaz let Strong in—only through a sliver of Mexican territory well to the west. Even so, sensitivities ran high. A Mexican publication, *El Monitor del Comercio*, expressed outrage when it learned that the Santa Fe would skip the original Guaymas and create a "New Guaymas" port of its own creation two miles away. Even more irritating to the newspaper editors was the revelation that the Santa Fe was naming the streets of this New Guaymas after its employees. After a San Francisco–based paper, the *Alta California*, rose to the Santa Fe's defense, the editor of the *El Monitor* responded in such viciously personal terms that the *Alta California*'s editor challenged him to a duel. The *El Monitor* editor sensibly declined, but rumors flew that the Americans were going to use their new railroad to invade militarily—provoking angry crowds to threaten to pull up some of its tracks in protest.

Díaz diffused the controversy by blaming it on disgruntled labor contractors and argued that the Santa Fe's New Guaymas would be a welcome addition since countless homes would otherwise have to be destroyed to bring tracks into the old one. For his part, Strong declared his respect for the "integrity" of the Mexican government and decreed that anyone who heard an employee of the Sonora Railroad speaking ill of the Mexican Republic to "notify Mr. Morley at once and he will discharge them."

By the terms of the agreement, the Santa Fe had fourteen months to go down the first sixty miles from Nogales, and twenty months for the rest. While A. A. Robinson was officially in charge, the bulk of the job fell to Morley, who was now getting worn down from his years of hard service. "Guaymas has a very pretty little harbor," he told his wife, Ada, back in New Mexico, "but the country around is not much. Quite barren and I think very hot in summer although probably not so bad as in the country." To secure the ranchland needed for the route, he plied the ranchers with liquor and then took time to "talk and smoke awhile" with them, Robinson noted approvingly. More often than not, he emerged with signed papers, and a "life-long friend." It was tough going, though. "I will tell you dear," he wrote Ada, "it is a job to go to a foreign country—and a semi-barbarous one at that and start building a railroad and have charge of and be responsible for everything." He wished he had never told Strong he'd do it. To close a letter to his wife on a cheerful note, he praised the local oysters. "I am chuck full of oysters all the time," he assured her. To cheer him up, Ada sent him his loyal horse King William.

As work proceeded, Morley's labor crews began to need police protection, and the Apache started stealing wire off the company's telegraph poles. Still, he forged on and laid the final rail into Nogales in October 1882, well ahead of schedule. Stretching back over seventeen hundred miles, it made for the longest stretch of track controlled by a single company anywhere in the world. The combination also created the nation's first international line. Robinson marked the occasion by bringing up one train from the south and another down from the north to touch cowcatchers in greeting. The Santa Fe stock price jumped when the news came in. It looked for all the world like Strong had broken Southern Pacific's monopoly on the Pacific, winning him prized entry into California at last.

CHAPTER 23

"This Is Hard"

BACK IN COLORADO, NEWS OF STRONG'S SUCCESSFUL drive to Guaymas led the General to make a rash decision. He'd build into Mexico himself, but not to the Pacific. While that had long been his quest, there was no point now that Strong was doing it—besides, the Treaty of Boston forbade him from building a connecting Rio Grande line south from Colorado to make any immediate use of it. No, he would start a whole new railroad, the Mexican National, to achieve something so grand that it would make Strong's accomplishment look puny. He wouldn't just gain a port. He'd take the whole country.

It was a lofty goal, but he saw clearly how to do it: He'd send a line down from the Texas border at El Paso all the way to Mexico City, and then run branch lines in either direction to hit the Gulf on one side, and the Pacific on the other. The country was big enough, and his plan vast enough, that the lines could be self-sustaining even without a link to the Rio Grande north of the border, which by the terms of the Treaty could not come down for another ten years. For now, he'd draw his traffic from the other railroad lines that had already reached El Paso.

He'd been considering a line into Mexico for almost a decade—but as all too often happened with Palmer's grand plans, he had not managed to act on it until he felt Strong had goaded him into it. Now, Palmer realized that the country presented another, bigger Colorado, and he would not content himself with just a sliver but take all of it.

The General had first visited Mexico in 1872 at the behest of Gen. William Rosecrans, his former military commander in Tennessee, who after the war had become the Mexican liaison for American train men seeking to do business in the country. Palmer had asked Queen to join him for the trip even though she was six months pregnant with Elsie, and she had agreed to come so long as she could bring Rose Kingsley. As a precaution, the General had taught the two of them the art of pistol shooting on the deck of a steamer on their way down from San Francisco. (To get to the steamer, they'd taken a Central Pacific train from Cheyenne, blasting through some heavy midwinter snow.)

The Palmer party landed at Manzanillo, well past Guaymas, below the Gulf of California on the Pacific coast. They immediately headed inland by carriage through the steamy jungle to Mexico City, where Palmer planned to make his case with the Mexican government. They'd be passing through the hazardous rebel country of Porfirio Díaz, then a fearsome revolutionary who was scheming to overthrow the elected president, and Palmer thought it might be helpful to have another gunman along. So, back on board the steamer, he'd invited a rugged-looking Mexican fellow to join them for the land journey. Palmer didn't realize that the man was in fact Porfirio Díaz himself, traveling incognito. Díaz rode with Palmer only a few miles, but had his rebels shadow the party as they rode on from there. One morning, a pair of his soldiers burst into the small hotel where the Palmer party was having breakfast to seize the guns that Palmer had brought along for his group's protection. No one was harmed, but the incident left poor Queen in paroxysms.

The group reached Mexico City safely, but when Palmer gathered a couple of allies to venture out from there to conduct a preliminary survey for his train route, their stage was ambushed by a gang of bandits, who let loose pistol fire. Palmer had by then acquired fresh arms, and he and his men shot back to repel the bandits, but not before Palmer's forearm was grazed by a bullet. He never breathed a word of his close call to Queen, and she never noticed the wound. His overture to the Mexican government went nowhere.

Undeterred by this experience, the General now, in 1880, settled on his audacious plan to run one train down from El Paso to Mexico City and add two others to reach the Gulf of Mexico and the Gulf of Calfornia from there.

As was their way, when Strong learned of Palmer's intentions to crisscross Mexico with trains, he decided to make the idea his own and do it with his usual alacrity. He soon discovered that a Mexican company, the Central Mexican, had secured the rights to build down to Mexico City from El Paso, but had been unable to assemble the funds to build. Strong snapped the rights up, inverted the name, and incorporated this new venture in Massachusetts as the Mexican Central, a wholly owned subsidiary of the Santa Fe. All he needed now was Díaz's blessing.

When Palmer heard the news, he nearly went insane. He raced to New York to meet with editors of the *World*—which had taken a special interest in the railroads since Tom Scott bought the paper in 1876—to try to kill off this latest and most dire challenge from Strong. This time, he would have to deliver a response that once would have been far beneath him. But not after Raton, after the Royal Gorge, and now after this. He claimed that Strong had no legal right to the charter, that he was facing a legal injunction, and that his own plans for the line were far more popular with Mexicans. None of this had an iota of truth. To hide his involvement, he sourced the assertions to the paper's Mexican City correspondent.

Strong forcefully rebutted the charges in his favored paper, the *Boston Evening Transcript*, declaring that these claims were not just false, but they were clearly not the product of anyone in Mexico, since they drew on reports that could not have reached New York so quickly on their own. He noted that Palmer was known to be in New York, and so "attacks upon the charter and concessions of the Mexican Central R. Co. Limited might be expected." The only fault in the matter, said the *Transcript,* should go to the "parties who had a former charter and for-

feited it and failed to secure a second recognition of the Government."
And who might they be? The "Palmer people," the paper flatly declared.

Caught out by the *Transcript*, Palmer directed his charges directly
to the subscribers of the Santa Fe's latest stock offering, in hopes they'd
now dump their shares and ruin the Santa Fe's plans. But the subscribers held firm.

Palmer's last, desperate hope was to rush to Mexico City and personally plead with Díaz to award him the El Paso concession. On September 8, 1880, Díaz gave his answer. No. He would award it to Strong.
He had amply demonstrated his bona fides by building down to Guaymas after all. For all his talk, what had Palmer ever done in Mexico?
Five days later, he threw Palmer a bone: a far less desirable route south
to Mexico City starting at out-of-the-way Laredo on the Texas border

Porfirio Díaz in full military regalia.

well down toward the Gulf from El Paso. He could build down from there, and good luck. The General lost Raton by two hours, the Royal Gorge by thirty minutes, and now El Paso by five days.

>─+─◦─◦+─<

Once Strong started building, his construction crews sailed down Mexico as if blown by the wind. "A man can build a railroad with red ants if he has enough of them and can keep them at it," the railroad writer Cy Warman once observed. And Strong had plenty of ants. At the height of the building, he had forty-five thousand men in his employ, five thousand more than the United States Army. The hard part was getting them all their pay, since they wanted it in silver. It took ten wagons, guarded by a well-armed cavalry, to come down from Texas banks and then fan out to the crews that were scattered all over Mexico.

The logistics were daunting, but not as bad as they'd been at Guaymas. Strong had to bring in much of his material—rails and locomotives, primarily—from England only to Veracruz on the Gulf of Mexico, not all the way around the Horn. And, to bring it on for distribution in Mexico City, he'd arranged for it to be loaded first on the Mexican Railway, before Palmer's. El Paso made for an easy starting point for Strong, too, since American trains could deliver all his materials there, and also benefited El Paso in return—by making it the major crossing point to Mexico. He boosted the fledgling city's population of barely two thousand in 1880 to almost twenty-five thousand by the end of the century.

Morley was in charge of building down to Mexico City, and it was on that route that he reached his end. Boarding a horse-drawn carriage to check out some of the construction, he'd taken a front seat with the driver while other railroad surveyors rode in the back. To make room for Morley, the driver had to move his rifle. It must have gotten tangled up in the reins, because it went off, blasting Morley in the chest. He sat there stricken for a moment, blood pouring from him. "I'm a dead man," he said. Then he stumbled down off the carriage and fell to the

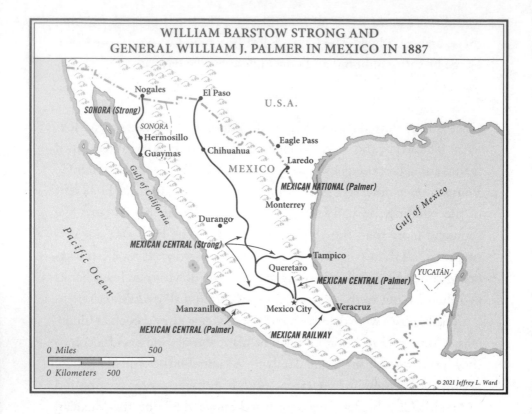

WILLIAM BARSTOW STRONG AND
GENERAL WILLIAM J. PALMER IN MEXICO IN 1887

ground. "Boys, I am sorry this has happened," he told his men gathered in shock around him. "This is hard." His last words.

The company brought Morley's body back by train. Morley's widow retrieved King William, too, and saw to it the horse was never ridden again. For the rest of his days, the horse lived on, she said, "in the best horse heaven which devotion could provide."

Despite the tragedy, Strong kept going. He completed the line from El Paso to Mexico City on March 8, 1884, having set the world record for tracklaying: 525 miles of track in 365 days.

CHAPTER 24

The A&P

PALMER COULD AFFORD TO DEPLOY ONLY A THIRD AS many ants as Strong, and the result was a disaster. The burning desert plateau on his route from Loredo was gashed every half mile with steep chasms, each one a pair of high, jagged rocks angling down to an abyss, "possibly the greatest obstacles of any road ever built," in the judgment of the *Railway Review*. To cross them, Palmer had to use countless switchbacks, tunnels, ridge cuts, and bridges, all of them time-consuming, expensive, and maddeningly difficult to construct. And, while Strong had no problems getting going at El Paso, Palmer got snarled in diplomatic red tape at Laredo. He was obliged to acquire formal clearance from Texas, the US Congress, and the secretary of state before he could cross the Rio Grande into Mexico, a process that took months. And while Strong's deliveries at Veracruz went straight onto the Mexican Railway trains to Mexico City, Palmer's rusted there until all of Strong's had been loaded. To get materials to Manzanillo, Palmer had to ship them around the Horn as Strong had at Guaymas, but the General then had to climb up ten thousand feet, requiring innumerable switchbacks and a 660-foot tunnel.

Desperate to make progress, Palmer worked three stretches of the line with five construction crews, which only exacerbated the logistical problems. After a full year, Palmer had come down from Laredo only twenty-five miles; a year and a half later, he'd reached only a quarter of the way to Mexico City. And his money troubles were mounting. He had relied on a Scottish syndicate, a London investment house, and a pool of small New York investors for funding, but the last batch of money was blocked when, in February 1884, William Rensens of New York City sued the Rio Grande for failing to live up to its financial presentations. Palmer was forced to admit that the company was running in the red, hurting his chances for further investment.

It got worse. Palmer banked with the Monte de Piedad—a former pawn shop that had gotten into banking only a few years before. It collapsed, taking the Rio Grande's operating funds in Mexico with it. With that, the line down from Laredo stalled out, and the one east from Manzanillo did, too. Neither would ever reach Mexico City. The General had tried his best, and he had failed utterly.

><-•>-0-<•><-<

Strong was not a man to take pleasure from Palmer's defeat. Besides, he had his own problems in Mexico. His route to Guaymas was proving a dismal failure of its own. Although the company stock had risen on completion, it started to fall shortly after, and by January 1883, it was clear to investors the thrill was gone. Huntington had been right. A run to the Pacific in Mexico was a fool's errand; Mexico was simply too foreign for American businessmen. The Sonora Railway was losing money hand over fist. In the 1883 annual report to investors, Strong tried to blame the slowdown on a fever epidemic in Mexico.

Despite these setbacks, Strong was not done. He was still convinced that California was his destiny—the culmination, the final shore—and he'd get there if it killed him. When the bad news from Guaymas was starting to come in, he turned to his fourth and final plan. Call it the backup of the backup of the backup, but it was the plan on which he had been pinning his greatest hopes from the beginning.

He would ride his own tracks to the state, not the California Southern's and not the Southern Pacific's. He would not rely on the kindness of others or some outlandish scheme like the Sonora Railway, but go straight west to California, hewing close to the line the Santa Fe had taken from Kansas, as he originally intended before he tangled with

Palmer. Relying on fresh tracks across barren, unpopulated desert, it was, in its way, the most audacious plan of all, but, to Strong, that meant that he was the man most likely to achieve it.

He'd started plotting the route the moment he got over Raton into New Mexico back in the summer of 1879. He'd put poor Morley on the job, along with another surveyor named Lewis Kingman. The route followed the 35th parallel that Palmer had been exploring when he ran afoul of the Apache back after first seeing Raton with Bell in 1868—which would certainly have vexed the General if he'd known. It ran halfway between the southerly 32nd that Huntington had taken through Yuma and the northerly 39th taken by the Union Pacific— and, coming across New Mexico, skirted the Rockies to the south. It ran on a line to hit California well below San Francisco in the state's relatively unpopulated south. Strong saw that region of the state as a more appealing version of New Mexico, especially the honey spot where the 35th parallel would lead—to Los Angeles. It was only a third the size of Denver, but to Strong this only meant it had plenty of room to grow. Besides, as a midwesterner from chilly Wisconsin, he saw Los Angeles as a place of eternal sunshine, a paradise by the sea.

In fact, the 35th parallel had been under consideration for a trans-continental route well before Palmer looked into it. In 1853, the federal government had sent out an army surveyor, Lieutenant Amiel Weeks Whipple, with a ten-man entourage that included a geologist and a naturalist *cum* artist who'd done work for the Prussian explorer Baron Alexander von Humboldt, to check it out. The team got so lost when its magnetic instruments were thrown off by volcanic rock that all their mules starved to death, and the survey party nearly did, too. Nonetheless, Whipple discovered a straight shot from southwestern New Mexico to Needles, California (so named for its distinctive, near-vertical slabs of rock), just past the Colorado River that defines the border with Arizona. No cutting through the mountains, no angling along the edge of Mexico. Just straight west across the blistering Mojave Desert. In calculating the costs, though, Whipple made a computational error that doubled the price, which dropped it from federal consideration.

The Kansas Pacific never had much of a chance to act on Palmer's advice once he took the line's western ambitions for himself, but John C. Frémont, the explorer and presidential candidate, saw the possibilities. He organized a railroad company called the Atlantic and Pacific to use that line, and in 1866, he won a federal land grant to build it within ten years.

For all of his other skills, Frémont proved a lousy railroad builder, and the A&P didn't even come close, plunging into bankruptcy in 1875. Soon after, a brand-new midwestern line, the St. Louis–San Francisco, known as the "Frisco," came into existence largely to snap up the A&P for its land grants. Strong took notice, and in November of 1879, he persuaded Nickerson to get in on the deal so the Santa Fe could ride the A&P rails to the Pacific.

The idea was for the two companies, the Santa Fe and the Frisco, to divide the A&P between them. They'd share the SF's route from Kansas to Albuquerque. From there, the two companies would combine to build A&P tracks to Needles, tracks that both lines would be free to use. If all else failed, this would be Strong's avenue to California.

But only *to* California, not necessarily *into* California. While Whipple had plotted a route on to the coast, Huntington had craftily inserted into the 1866 legislation a provision that allowed his railroad the right to meet the A&P at Needles to carry the traffic into California from there. Since the Southern Pacific had already gone to the trouble and expense of building across California to go through Yuma, far to the south, to defy Tom Scott, Strong couldn't believe that Huntington would run another set of tracks to Needles just to keep him out. But this was 1879, long before history would teach him otherwise.

Strong had Robinson send out his new man Kingman to survey the new A&P line across Arizona. Though Kingman intended to maintain peaceful relations with the local Navajo, he came equipped with eight Colt revolvers, twelve carbines, and seven hundred cartridges, just in case. He quickly learned that in the broiling desert, dehydration proved the greater threat. Water was so scarce that Kingman had to limit each man to a single pint a day.

While Kingman relied on Whipple's initial survey, he made good use of Palmer's, too. Once he was ready to build, he picked tiny Isleta, a dozen miles south of Albuquerque as the departure point and brought in eight hundred boxcars' worth of rails and ties. Remembering the fights with the Rio Grande at Raton and the Royal Gorge, Kingman sent a full load of rails by wagon 180 miles ahead to a spot along the route through a narrow pass, where there would be room only for a single set of tracks. He laid those rails right away, a full year before they'd be reached, to secure the rights well in advance. For the work itself, Kingman relied on crews of Mormons led by a son of Brigham Young, plus Mojave Indians. It was grueling work in the scalding heat, but the crews managed to set down a mile of track a day—until they hit a wide, deep gap in the Arizona plateau called the Canyon Diablo, or Devil's Canyon, that plunged 250 feet down through shimmering limestone. In anticipation, Kingman had ordered from New York a prefabricated

The A&P bridge over Canyon Diablo.

bridgework, 560 feet long, to stretch out across the canyon on a scaffolding of angled trestles that were set down on the rocky floor. Once erected, it looked like a teetering wall of matchsticks, but it held, and the A&P raced on. Soon, Kingman hit another obstacle in the forbidding Johnson Canyon but was able to cross it by easing down one side and then driving a tunnel through the hardened volcanic lava on the other. Six workers were crushed to death in an accidental dynamite blast, but the tunnel proved a thing of beauty, a Gothic archway into the darkness. By September of 1880, Kingman had crossed the burning desert to the mountains that stood a cool, luscious green with pine and juniper that gave way to shimmering Aspens at the higher elevations.

▸⊷•⊶◦⊷•⊷◅

As the A&P forged ahead, Huntington followed Kingman's race to the California border like a pirate king gazing through a spyglass. Whatever Strong might have hoped, Huntington saw in him nothing but an enemy. To defeat him, Huntington would build out to the A&P at Needles; Strong's line would not come an inch into his state. To be on the safe side, he'd keep Strong from even getting that far. He'd thought of teaming up with Gould once before to take the A&P. Now he would do it. Gould had his own reasons to want this new western line gone: the A&P was threatening his own lines, the Union Pacific and Kansas Pacific, farther north. Huntington would team with Gould to buy out the Frisco's half of the A&P, stop it dead at Needles, and then turn that line against the Santa Fe's network in the Midwest. They'd tie Strong down and then choke him to death.

The deal was struck in January of 1882. That spring, Gould and Huntington used their control of the Frisco to stop the A&P tracks on the Arizona side of the Colorado River, banishing any notion that Strong might go farther. Lest there be any remaining doubt, Huntington built out his line to Needles down through the middle of the state and then set the price for using it impossibly high. As ever, Strong professed no public concern. By now, T. Jefferson Coolidge had resigned the presidency of the railroad—largely out of concern that Strong's relentless building would bleed the company dry and ruin his investment—and the Santa Fe board had replaced him with Strong, since he was the prime mover behind the railroad. The appointment made Strong the first president to be a manager only, without any significant investment in the company. It also placed the fate of the Santa Fe, at the most precarious point in its history, entirely in his hands.

Whatever Strong proclaimed publicly, as the summer of 1882 turned toward fall and Huntington and Gould blocked his last hope to enter California, Strong had to fear this was the end. His dreams were all dashed: He would never take the Santa Fe to the Pacific. It was not his destiny after all. It had to have been excruciating. For all his plans and backup plans, nothing had gone his way. It was crushing: Having been right about everything in his fight with Palmer, he was wrong about everything in his fight with Huntington. And the stakes were far higher now. Strong had imagined that the California Southern would get him to the California border. Wrong. He thought he could ride west into the Golden State on the Southern Pacific tracks. Wrong. He figured Guaymas would be his salvation, allowing him to breach the walls of Huntington's Fortress California with a sneak attack from the sea. Wrong. He was sure the A&P would come through for him. Wrong again. Wrong, wrong, wrong, wrong. He'd been wrong about all of it.

Always before, Strong had been able to recast any defeat as merely a temporary setback on his long march to triumph. Not this time. This felt all too sweeping and all too final. He'd not just lost but lost completely. He'd bet the company on his genius, and he'd failed. He'd believed that the Santa Fe had to Grow or Die. Now it had grown *and* died. His California dream was over, his career wrecked, and his railroad doomed.

<hr />

Until that grim fall of 1882, Strong was a happy man, merrily bounding from success to success. How did he feel to be brought so low? Miserable? Furious? Frantic? Or merely resigned?

Precious little evidence of Strong's inner life survives, but, as it happens, there are some personal letters from this period that detail his emotional ordeal. They start in 1881 with Strong's transition from Santa Fe's general manager in Topeka, to its president with a grand office at the company's stately corporate headquarters on Devonshire Street in downtown Boston. The first letter, in fact, refers to his move from his family home at 289 Erie Street in Chicago to a new address in Brookline, a pleasant, leafy town that had been carved out of Boston as a separate municipality of the well-to-do. That September, Strong had shipped east from Chicago all of his household goods aboard two thirty-three-foot boxcars. While the Big Four were competing to see which of their houses could reach highest to the sky, Strong described his new abode to his insurance company only as a "first class, modern frame house, about 100 yards away from any dwelling, in a city amply supplied with water from an efficient fire dept." He'd bought a neighbor's house on Erie Street as an investment, and now asked a Chicago realtor to put it up for sale, as well. "Now as to price," Strong wrote him, "I will not accept less than $20,000 *and want as much as I can get*." The fact that he had only such modest investments had emerged as a point of resentment when he took over the presidency from Coolidge, but he was careful that any investments he might make not conflict with the interests of the railroad. "Ever since the first day I commenced railroading," he informed a colleague, "I have never allowed myself to be invested personally in any company or corporation in which my official acts could have any bearing for or against my personal interests. I think that rule has saved me a great deal of trouble."

As his hopes for California faded the next year when he lost control of the A&P, Strong's mood darkened noticeably. He'd made C. C. Wheeler his first lieutenant in the westward expansion, and a rare confidant. Wheeler must have intimated his own distress to discover that Huntington and Gould were teaming up against the Santa Fe, for Strong wrote to him in the spring of 1882: "No wonder you are blue. There is reason for it." A few weeks later, Strong tried to buck his friend up. "I have never seen our people as determined as now," he assured him. "The situation is not as bad as it might appear."

Strong, however, was all too aware that the situation was dire. If he was ever to get into California, he needed control of "the Frisco road." Without it, he confided to Wheeler, "the scheme would prove a failure." Strong's purchase of a half share in the A&P had initially been "considered visionary," he reminded Wheeler, like his bold leasing of the Rio Grande from Palmer during his war for the gorge. But he had to admit the move did not seem so visionary now.

Once it was clear by midsummer that he'd lost control of the A&P, Strong turned unusually bitter. "My only regret or rather my greater regret is that the parties who made the blunder are now free from the care of handling the matter." He was referring to the nefarious Coolidge, but it was unfair to lay all the blame on him. Strong just needed a convenient place to direct his anger. "He is not considered as level headed a man as he was a year ago," he snapped.

Nevertheless, he told Wheeler he still held out hope of using the Southern Pacific rails, not just to California's border, but well inside. "Mohave is better," he told his colleague, referring to the town on the edge of the desert, just shy of Los Angeles. "SF better still." It was now a fantasy, but a fantasy was all he had. "A&P is moving on," he assured Wheeler. "Truce prevails till we get to the Col. River." Needles was the best he could genuinely hope for, but he had clung to the idea that he would beat Huntington there just as he'd ultimately defeated Palmer at his Dead-line. "There will commence tug of war," he admitted, but he evinced his characteristic optimism: "The A&P has surrendered nothing," he insisted. "Be ready to move, and in 'full blast.'"

Then he faltered. In late July, Strong fell sick with an unspecified "severe illness," so debilitating that his executive secretary, Smith, voiced his own anxieties about the situation. The trouble apparently involved Strong's vision. Smith was reluctant to write down the details but termed it "a very severe and protracted surgical operation," before admitting it was "about the eye," and it was not the first such operation Strong had undergone. He was encouraged that the surgeons were optimistic. "I hope and believe that all is well, but it is a trying time just now." Ten days later, Smith reported that Strong was improving, "but he is far from being out of danger." He was still having trouble with his eyesight, a grave matter for any executive. "He is still unable to use his eyes or his pen, and it is hardly to be hoped that he will be able to attend to business for a long time to come." It turned out, he was wrong about that, for Strong's vision returned in a matter of days. Just a week later, Smith declared his boss "out of danger."

To fully recover from the ordeal, Strong stayed out of the office the rest of the summer, putting out of his mind his mortal struggle with Huntington. Strong didn't return until September 13, and then just for one day a week. He confided to Wheeler that he'd gone through "a pretty tough siege," and even now was "doing very little." He insisted he was, at that point, more worried about Wheeler, who was reporting medical woes of his own. *Remember take care of your health*," he advised his friend, speaking from experience.

On November 11, Strong admitted that he had become extremely distressed about the company's financial position. The failure at Guaymas, compounded by the stall-out at Needles, meant that revenues "have not met my expectations," which was putting it mildly. The company was on the verge of bankruptcy. He had to face the fact that, despite all his efforts, he had been locked out of California. His railroad was dying.

Then came the bombshell. Strong told Wheeler he'd tendered his resignation to the board, which had responded by granting him a six-month leave. To Wheeler, he made it sound like the issue involved the company's precarious finances, but actually it was his mental health. Strong worried that his ally might be under the same emotional strain, and that he might be thrown by the full truth. Wheeler was set to take over for Strong as acting president, and Strong desperately needed him to keep things going in his absence. To ease his burden, Strong advised Wheeler to hire an executive assistant "as soon as you can, for I fear you will break down. Be careful."

A few hours later, Strong reconsidered, and revealed to his friend the facts of the matter. "I have been compelled to put up the white flag," he wrote him. "You will only know the struggle it has cost me, when, if you ever pass through the same experience. I pray you never have to know the trial." He enclosed a copy of the board's formal notice, granting Strong the six months off. "I shall take only a part of the time allowed but enough to put me on my feet," he assured Wheeler. "My head is sound but my nerves have nearly got the best of me, which is a new experience for me." Later that same day, he wrote Wheeler once more to let him know that the Santa Fe might never get into California. "[The prospect] has disappointed me, but it cannot be cared as that we have done our share—" He stopped, seemingly distracted, and then lurched on incoherently: "cuts off all criticism for none has been made." His final words reveal his true state of mind: "If you were here I would throw my arms around you." He was immensely relieved that Wheeler would carry on for him. "The sorry thought that you might leave gave me the blues." That is the last personal letter.

CHAPTER 25

▶─▪◦─◈─▮◦─◈─▮◦─◈─▮◦─◈─▮◦──◀◄

The End of the Line

RUINOUS AS PALMER'S MEXICAN VENTURE PROVED TO be, things got even worse for him north of the border. In April of 1880, shortly after he'd won his Leadville exclusive from Gould, the miners all went out on strike, demanding a twenty-five percent raise and insisting their safety be secured as well. When the owners refused to oblige, the men threatened to turn the so-called Magic City into Bloody Leadville. To deal with the unrest, Haw Tabor persuaded the governor to send out the state militia led by the Rocky Mountain Detective Association's Dave Cook, who also served as Denver's city marshal. Cook was evenhanded, dealing with the strikers and the strike breakers with equal firmness. He ended the debacle, but the resolution yielded no improvement in pay or working conditions. The dispute also took its toll on silver stocks, starting with Tabor's Little Pittsburg. The financial bloom was off the rose in Leadville.

Luckily for Tabor, his Matchless Mine continued to crank out silver at the extraordinary rate of $1 million a year, setting him on a course to become the richest man on earth. To get clear of Leadville's declining fortunes, he moved to Denver, where he created yet another opera house, this one a five-story extravaganza called the Tabor Grand Opera House, its vast interior hall lit by an immense crystal chandelier bearing hundreds of gas jets to produce a veritable sun. Although the *Rocky Mountain News* denounced Tabor as a "shambling, illiterate boor," he persuaded the governor to appoint him to fill out the remainder of the term of a Colorado senator who'd died in office. That lasted only a month, but Tabor used the time to wed Baby Doe as a US senator at the capital's glittering Willard Hotel, with President Chester A. Arthur in attendance. As a wedding present, Haw bought Baby Doe a jeweled necklace that was said to have been sold by Queen Isabella of Spain to finance Columbus's voyage to America, but probably was not.

Distraught over her husband's nuptials, ex-wife Augusta Tabor took a whack at Haw's character with unflattering revelations about him. When Baby Doe got Haw to drop his business manager for her brother,

The Tabor Grand Opera House.

the business manager retaliated by doing his best to finish Haw's reputation off. In mounting distress, Tabor made increasingly improbable
investments, but now his luck failed him; none panned out. His final
undoing came in 1893, when Congress removed silver as a backing for
the dollar, crashing the price of silver. Tabor dropped into bankruptcy,
and he was reduced to working as a common laborer in Leadville before
some friends got him appointed Denver postmaster. After Tabor died
in 1899, Baby Doe clung to the idea that her silver fortune someday
would be restored. She held on to Haw's Matchless Mine, moved into
the shack next door, and waited for the happy day that never came.
She died there one winter decades later. Her body was found frozen on
the floor.

Seeing Leadville's prospects dim, Palmer looked for other mining
towns to restore him, sending a spur deep into the San Juan mountains
past Alamosa to the west. It was a venture born of desperation. To one
newspaper correspondent, Palmer's line had become a "knight-errant
railway in quest of adventures, a New Columbus, with cars instead of
ships, in search of undiscovered realms." Columbus, of course, found
something; Palmer did not. In going into the torturous mountains
past San Juan, the railroad faced engineering difficulties that made
the challenge addressed by McMurtrie's Mule Shoe Curve look like a
snap. For one stretch, the mountainside required so many switchbacks,
Palmer had to lay a full two and a half miles of track to progress a single
half mile. Even worse, Palmer had to pay top rates to find enough
workmen for all the frantic tracklaying. The idea was to cash in on the
real estate along the route, and feast on some silver mines. But none
of it paid off.

In a panic, Palmer decided to build on past Leadville, through the
Rockies, and across the Utah desert to Ogden to connect with the Central Pacific and reach California that way. Forget the southern route
to Mexico he'd always planned on. In fact, forget all of Mexico, even

though he had over a thousand men laboring there. Mexico would never save him now. It was California or bust. To go to California via Utah, though, he had to stay ahead of Jay Gould. So Palmer threw his Rio Grande over the mountains, crossing the sky-high Marshall Pass to Gunnison before Gould could make a move. But such haste made terrible waste. In his rush, he'd sent his construction crews ahead of the surveyors plotting the route, requiring miles of track to be ripped up and relaid once surveyors determined the right way to go. In his mounting anger over such extra expenses, Palmer blamed his ingenious engineer McMurtrie and his loyal treasurer, Weitbrec, but the mistakes were completely of his own making.

In reality, Gould had never thought of trying to claim that route; he just wanted to bluff Palmer into a spasm of fruitless tracklaying. Gould was still a shareholder in the Rio Grande, and now, he thought he might want to lower the stock price and raise his stake. Even if he didn't, he found he rather enjoyed the spectacle of bringing the imperious Palmer down. To lower the price further, he dropped hints of the dreadful state of the Rio Grande's finances to the press. Palmer tried furiously to fight off the allegations, going so far as to sue the *Wall Street Daily News* for libel, but there was too much truth to them. To save himself, Palmer had the ever-loyal Willie Bell start an allied railroad, the Denver & Rio Grande Western Railway Company (or the Western, for short) that was free of the financial strain on Palmer's own Rio Grande, to build east toward Colorado from Salt Lake City, freeing the Rio Grande to build only halfway.

Palmer saw Utah as another virgin Colorado, and he somehow scrounged up the money to throw a thousand workmen at the task of laying track through the mountains. A few days before Christmas 1882, he reached the halfway point at the Utah line where Willie Bell drove the Western crews hard to meet him. Annoyed to see Palmer's combined line enter Ogden, Gould tried to ban his train from the Union

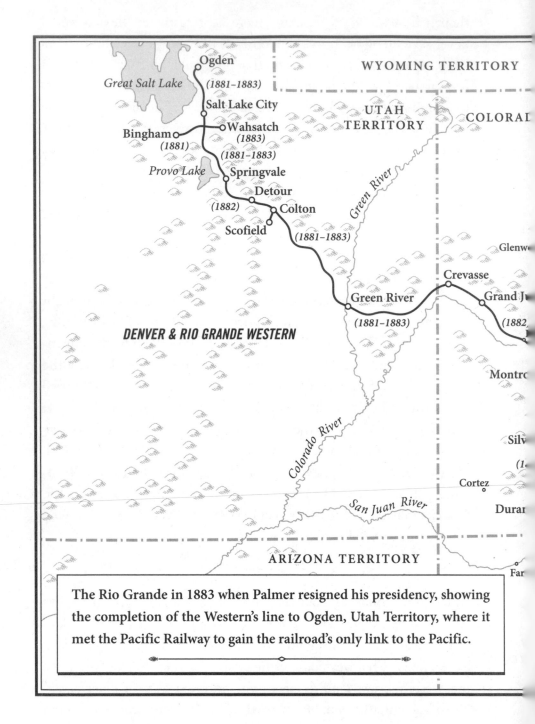

The Rio Grande in 1883 when Palmer resigned his presidency, showing the completion of the Western's line to Ogden, Utah Territory, where it met the Pacific Railway to gain the railroad's only link to the Pacific.

THE RIO GRANDE'S ROUTES THROUGH THE ROCKIES AND INTO UTAH, 1883

0 Miles 50 100

0 Kilometers 100

Steamboat Springs

Kremmling

Orestod

Dotsero

Minturn Rock Creek

ngs

Fort Logan Denver

Dillon

Castle Rock
(1880)

Wheeler
Kokomo O'Briens Quarry

Malta Leadville
(1871–1872)

Aspen

Anthrocite *(1880)*

Manitou
(1880) Colorado Springs

set

rested Butte *(1881)* *(1882)* Calumet

Gunnison Monarch Salida Cañon City *(1871–1872)*

pinero *(1881)* *(1880)* Florence

(882) Poncha Jct. Orient Coal Mine Pueblo
(1881) *(1872)*

gway

Lake City Villa Grove *(1876)* DENVER & RIO GRANDE

Creede

Wagon Wheel Gap Alamo Cuchara

South Fork Francisco *(1876)*

kwood *(1878)* *(1877)* Rouse Junction

Alamosa Laveta

Pagosa Springs Russell El Moro

(1880) Fort Garland Trinidad *(1879)*

(1880) Antonito Coal Mine

Pagosa Junction

Lumberton Chama Raton INDIAN TERRITORY (OKLAHOMA)

Tierra Amarillo *(1880)*

La Madera

Gallina Caliente

Española TEXAS

MEXICO
RRITORY

Santa Fe

© 2021 Jeffrey L. Ward

Pacific's station, but Palmer skirted the issue by going in at the Central Pacific's end.

Palmer had joined with the Western by leasing it, but that investment did not sit well with the Rio Grande bondholders, who worried that the railroad was already dangerously overextended. Happily for the RG investors, Palmer's failing Mexican National stood as an entirely separate entity, but the Mexico disaster weighed on him. Even though it had successfully reached Utah, the Rio Grande was in terrible financial shape—and it didn't seem like Palmer was the man to make it better. When some of the board members turned on him, Willie Bell and a few other Palmer loyalists grew indignant that anyone question the heroic founder's leadership? When their encomiums were received with icy indifference, Bell and the others resigned in outrage. But that only meant that the General's friends on the board were replaced with foes, further weakening his position. The newly constituted board shifted the Rio Grande's headquarters from Colorado Springs to New York City, revealing a marked shift in priorities, and made pointed inquiries into Palmer's wild spending. Refusing to concede an iota of fault, Palmer turned on his assailants, calling their interests in his railroad "purely wild and speculative," and went directly after one of them for his "reckless improvidence." On August 9, 1883, some directors laced into Palmer with such vituperation that he stormed out of the board room in a rage and came back with a hastily scrawled notice of resignation. The board accepted it. That was the end of the General's presidency of the Rio Grande. After twelve years, and more than twelve hundred miles of track, much of it through the twisting mountain passages of the Rockies, and the creation of dozens of new Western towns, Colorado Springs not the least of them, Palmer was finished with his baby railroad.

The Rio Grande board replaced the General with a man he detested, Frederick Lovejoy, a Philadelphian with no experience in railroading beyond an investment in the Pennsylvania, revealing that his interest in the Rio Grande was vulgarly pecuniary.

By then, the Rio Grande finances were in such desperate straits that Lovejoy's first move was to secure an emergency $50 million mortgage, half of it, he said, "to provide for present needs." At that, Gould moved in for the kill. Having no further use for the Rio Grande, he resolved to finish it off. He bought a competing line to dig into the San Juan coal fields that the Rio Grande depended on, and cut the UP's rates into Ogden, forcing the Western to match his price. Meanwhile, Lovejoy sold off Palmer's ancillary Colorado Coal and Iron Company, removing the last of his predecessor's interests from the Rio Grande. He also installed his own manager of the Western and ousted the few remaining Palmerites from the Rio Grande board.

Still, Palmer remained president of the Western, and in that position he sued his old company for abusing his new one. The lessee turned on his lessor, producing a relationship the *Chicago Times* called "one of the most peculiar on record," although, of course, it had ample precedent in Palmer's struggle with Strong when the two contended for the Royal Gorge.

In retaliation, Lovejoy claimed that Palmer had defrauded him with the Western lease. Palmer replied that any fraud originated with Lovejoy's board. Lawsuits flew, and both railroads atrophied. When the court sided with the Western, Lovejoy cut it free. Literally—in his fury, he had men pull up a full mile of Rio Grande track where they joined the Western.

As the Rio Grande dropped into insolvency, bondholders demanded that the courts put it in receivership. Since his Western was no better off, Palmer filed for receivership, too. Angered by their short pay, five hundred Rio Grande employees went out on strike. When Lovejoy put in scab replacements, the strikers dynamited the Rio Grande tracks. Even after the strike was resolved, the Rio Grande was all but dead. To save the company, a reorganization committee proposed a reunification with the Western, but Palmer refused to take his baby railroad back.

On July 12, 1886, Lovejoy had no choice but to put the Rio Grande up for sale. It was bought for pennies on the dollar by a group of English and American investors who renamed the railroad the Denver & Rio Grande *Railroad*, not Railway. The Western and the Rio Grande both limped on, drastically harmed by the debacle. Their glory years were over.

CHAPTER 26

Ightham Mote

BY THEN, LIKE HIS RAILROADING, PALMER'S MARRIAGE
was finished in all but name, too.

After her heart attack, and the birth of their second child, Dorothy,
in 1880, Queenie had refused to return west. In a last desperate effort
to make Glen Eyrie into a house his wife could bear to share with him,
Palmer set about making it even larger, more extravagant and more
European than before. During the renovations he suggested to Queen
that they travel to England in a last attempt to stay together. Palmer
had been planning to go there to take his late father-in-law's three boys
to attend Oxford, anyway. Queen accepted and brought along their two
children—Elsie now nine and Dorothy just one.

Pregnant again, Queen gave birth to their third daughter—named
Marjory until they could agree on something better (which they ultimately could not)—in England in 1881. When their sojourn there was
complete, Queen told Palmer that she would not return to America
with him. Although the rebuilding of Glen Eyrie was complete, Palmer
went back to Colorado Springs to live there all alone.

On the return passage, he struck up an unlikely friendship with
none other than Oscar Wilde, bound for America on a lecture tour. "A
young Englishman," Palmer called him, "with a smooth, boyish face,
rather pale; a large, good eye, long hair and a mild expression which
lit up a good deal in talking; very good-natured, rather sunny in fact."
After paying this exotic close attention, the General pronounced him-

The three Palmer sisters, Elsie at nine holding
baby Marjory, and Dorothy standing beside.

self "pretty well acquainted" with Wilde by the time the trip was over.
It is hard to know if, to the General, Wilde's homosexuality was un-
noticeable, overlooked, or part of his appeal, but Palmer was plainly
not put off by this peacock whom he pronounced "the soul of the den."

As the ship entered New York Harbor Palmer wrote a letter to
Queen that detailed his torture. He professed to be full of longing for
the family he'd deserted. "My little family is the oasis," he wrote, "the
only green spot to which my mind can turn without distress & disappointment." He invoked his girls. "The Gentle Elsie on her Pony
- the Sagacious Eager little Dorothy. And the innocent sleeper in the
cradle." Then he turned his thoughts to Queenie. "And you my dear
wife whom I have so often cruelly distressed & whose young affections
I allowed to be Estranged because I was hard & cold & blind and stupid
& wretchedly wrong altogether, and Reckless. What a heaven would
life seem now, if with vigorous health, one had nothing to do but start
without a penny to make a home for this beloved flock."

It would be two years, in the spring of 1883, before Palmer crossed the Atlantic once more to see his family in England. When the visit was over, he insisted on drawing everyone back with him to Colorado Springs for one last try. Reluctantly, Queen agreed, but it didn't work. Queen had scarcely arrived before she declared her weak heart couldn't take it. She took their three daughters east to the original Newport, the posh resort of the moneyed crowd in Rhode Island. From there, she moved on to rent a lavish apartment in the new Dakota building in Manhattan, just west of Central Park.* With the interior done in an exuberant mélange of European styles—French Renaissance, Victorian English, German Gothic—it was several cuts above even the latest incarnation of Glen Eyrie. And it featured modern conveniences like elevators and central heating, the latter of which must have seemed heaven sent.

But New York proved to have brutal winters of its own. The first one dumped four feet of snow on the city and delivered a bronchial infection that drove Queen to her bed for months. In response, Queen gave up on America altogether. She took the children to England once more, and this time for good, settling in a medieval manor house well outside of London.

>⦁⊙⦁<

Ightham Mote (pronounced by the English "item moot") had been built by Saxons in the fourteenth century and done in rough stone with Tudor-style timbering. If Glen Eyrie was ersatz, Ightham Mote was the real thing. The interior surrounded a great hall with an immense arched ceiling, forty feet high, the space dimly lit in the medieval manner by the small, mullioned windows or flickering candles. A heavy Jacobean staircase rose past newel posts sculpted into Saracen heads to a second floor that featured a barrel-vaulted chapel amid the bedrooms, complete with a claviorganum, part harpsichord, part organ. The whole property had a mystical arrangement, as its central court-

---

* It was called the Dakota because as an early building on New York's emerging Upper West Side it was considered to be on the western frontier.

yard was comprised of widening concentric squares that ended in high walls over the swift, green waters of a surrounding moat.

Perpetually cold and damp, Ightham Mote was a place of some gloom, but to Queen it was everything Glen Eyrie was not. She made it a cultural retreat, drawing the eminent aesthetes of the day like Henry James, John Singer Sargent, and, yes, the ubiquitous Oscar Wilde, who were enchanted by Gothic rumors that the Mote was haunted by an early owner who'd been walled up behind the chimney.

Of all the ghosts that flitted about, the most haunting was the spirit of the General, now permanently ensconced in his mountain aerie across the sea. His absence particularly troubled Elsie, and she wrote a short story after she arrived that alluded to his disappearance from her life. "Man Overboard" it was called, and it described an incident from a recent passage aboard ship to England. After a passenger jumped overboard in a suicide attempt, the captain turned the ship around and sent out a lifeboat to rescue him. But after the man was brought safely back on board, he had to be restrained from plunging into the sea once more.

The Palmer sisters on horseback at Ightham Mote.

When Queen reported to the General back in Colorado that Elsie was looking "dull-eyed and wan," he recommended a regimen of vigorous exercise. He projected his own misery onto his daughter as he decried the "hot-house tendency" that brought the "competition of the business and social world" into the nursery. He didn't want to make "a little machine of her"—like the bigger one he'd become. When he wrote Elsie directly, he composed for her a childish ditty that recalled their happy days together at Glen Eyrie and carried it on for a dozen lines before abandoning it abruptly to turn darkly confessional: "I am all alone again."

To complete this sad tableau, Queen was so sure she was dying that she made young Elsie, now fifteen, promise to raise her sisters as their mother had once she herself was gone. "My big darling—my white little maid—my gentle snow drop—Elsie—you know how much we have talked of your being the little mother of the other two," she begged her daughter. "Make them brave and good—kind and true . . . and let them know how 'mutterlein' is loving them—and near them." On she went, pressing this loving hand ever more heavily until Elsie must have nearly screamed. However pressured it was for Elsie, it was feather light for her mother; she soon went bustling off on a shopping tour of the Continent, leaving all three daughters in the care of a Germanic governess she'd brought back from the Rockies. That left poor Elsie to write plaintively to her vacationing Motherling from the Mote, "Here I am, writing to you, my Mother, in your morning room which shows a tiny bit of your spirit to me . . . loving you dearly, dearly, and wondering what you may be doing just now at this very minute."

In 1887, Queen told Palmer that she would like to make the separation official. Not a divorce, but not a marriage either. She tried to couch this as a matter of health, claiming she couldn't take high-altitude Colorado, but then made a mockery of the argument, and of Palmer, by concluding she was now off to the Alps with some friends.

John Singer Sargent captured the terrible strain of the household triangulation on Elsie in an oil portrait of the seventeen-year-old girl, which he titled "Miss Elsie Palmer." It took a towering painting, a full six feet high, for Sargent to achieve the close scrutiny he envisioned. For dozens of mind- and body-numbing sittings, he had her pose in a stiff chair in the grim and drafty upstairs chapel, although there were plenty of sunnier places about. Ramrod straight, her hands clasped in her lap, she stared down her painter as if he was her executioner.

Portrait of Miss Elsie Palmer by John Singer Sargent.

In the finished portrait, Elsie resembled a ghoul from a ghost story by Henry James. Her skin was wan and nearly lifeless against virgin-white silk, her long hair parted perfectly on either side of her face, her lips slightly reddened with lipstick that hints at sexual desires that might never come. Her father's steely sense of purpose is visible in her strict verticality, her mother's flighty evasions in all the puffy satin that surrounds her. It's painful to behold, this upright corpse, her inner self hidden behind her outer one, her whole body in tortured symmetry as if pulled by precisely equal, precisely opposite forces, her whole being balanced on the tip of one big toe that descends down the axis of the portrait to touch the floor at a pinpoint.

Could this possibly be Miss Elsie Palmer? Elsie herself had no idea. In 1891 she confided to her diary: "There is . . . something strangely lacking in me . . . I feel the lack of it (whatever it is) must be at the room [sic] of my clumsiness, forgetfulness, shyness, vagueness, and it makes me feel—well, not very happy." She closed: "Good-night, feelings, happy, sad and complicated! I have had quite enough of you."

⊷⊶⊷○⊶⊷⊷

But they had not had enough of her. Queen's English friend Alma Strettell, who'd accompanied her to Leadville on the day of her heart attack, married a much younger man, Peter Harrison, a sly, elongated figure with a Van Dyke beard and a priapic streak. Something of an artist himself, he was a friend of Sargent's and had been fascinated by the "very calm expression" on Elsie's face in her portrait. He was determined to know what lay behind it, scandalously so. He wooed the girl ardently, scarcely pausing when Queen finally did die shortly after Christmas 1894.

She died as she lived, in a fit of histrionics, suffering from an asthma attack that left her so incapacitated that Elsie was summoned to attend her—but was forbidden from actually seeing her. She was instructed to communicate with her mother only by letter. From downstairs, Elsie's last words to her mother were in a thank-you note for her Christmas gift of a dress. She assured her that she and her sisters, all of them gathered below, "[feel] your presence with us every second." By then, fearing the worse, Elsie had cabled her father to come immediately, but he arrived far too late. Queen died the next day.

That was when, at Harrison's urging, Elsie allowed him to discover what lay within her. "Do wear it," he wrote her in one of his many passionate letters, referring to that very expression Sargent had captured. "When it comes on your face, your lips seem like folded hand & I love it & rest in it. My dearest, I love & love you—let me say it." When Elsie's sister Marjory discovered the attachment, she tried to displace her in his affections, only for Peter to fall for the third sister, nicknamed Dos, instead. And on it went with Harrison, all three sisters whirling about him, some of this three-way courtship spun out on moonlit nights in Colorado, where their father had brought them after Queen's death, right under the General's nose. He was so unsuspecting that when Harrison painted portraits of the three sisters, Palmer framed them for his Glen Eyrie wall.

Eventually, the whole absurd romantic contraption collapsed under its own strain, and Elsie gave up on Harrison to marry an odd character her father had introduced her to, years before. Seven years younger than Elsie, Leo Myers had devoted himself to the paranormal interests of his late father and was inclined toward the dreary. The marriage was not a happy one, not least because the couple settled by the Thames in London, just down the road from Peter Harrison and Alma Strettell. In time, Myers weaned himself from the occult and used his inheritance to indulge himself in writing obscure novels which, eerily, took on some of the General's dispositions. In a late volume, *The Root and the Flower*, Myers described an older man who might have been the General musing in a flower garden that might have been at Glen Eyrie. "The world seemed to him a place of extraordinary beauty," Myers wrote. Everyone should retreat to a garden. Then "all the earth would be yours to enjoy in a disinterested ravishment. The striving between man and man would have vanished. Paradise was as simple as that."

⊱─◌─◌─◌─⊰

Two years after his Queen passed from the earth, Palmer saw his beloved baby railroad fall into the hands of Jay Gould's son George. By
then, Palmer's tormentor had succumbed to the tuberculosis that had
been slowly consuming him. George had little of his father's prowess,
and the Rio Grande dwindled away, passing through a series of owners
in the next century.

The Western hung on, but Palmer's role increasingly became that
of a figurehead until 1901, when he was bought out for an impressive
$1 million. His own finances were well established by then through his
many ancillary investments in mines, real estate, and coal companies,
and he graciously distributed the money from the Western sale to all of
the railroad's employees who made up what was left of his "little family."
He devoted himself to local philanthropy, greatly expanding the school
he had founded, Colorado College, and adding to the town's parks and
nature preserves. In keeping with the abolitionism that pulled him into
the Civil War, he also donated substantial sums to an all-black college,
the Hampton Institute of Virginia.

In the hopes of luring his three daughters back to him, he redid
Glen Eyrie one last time, this time pushing the room count from
twenty-two to a staggering sixty-seven, raising the ceiling of the great
hall to twenty-five feet, installing Turkish baths and a rose arbor, and
adding a massive tower with a bell that could be heard six miles away.
He pronounced himself pleased. "It seems to be the sort of paradise to
which only the elect can be permitted to go."

>-+-0-+-<

In October 1906, Palmer was out on a morning gallop when the loose-
legged riding style he'd long favored finally caught up with him. His
horse stumbled, pitching the General forward onto the rocky ground,
where he landed headfirst. He snapped his spine at his neck. The fall
left him a quadriplegic: he was able only to rasp out speech, blink, and
shift his gaze, nothing more. In one photograph of the General in bed,
he seems to be a head only. After some survivors of the Fifteenth Pennsylvania Volunteers won him a Congressional Medal of Honor for his

Palmer in his Stanley Steamer automobile.

service in the war, Palmer summoned three hundred of his old regiment to Colorado Springs for a last assembly. A photograph shows Palmer in his wheelchair front and center of them. All the others are standing in evening clothes; the General sits in a stunning white suit, his Great Dane, Yorick, at attention beside him.

To fight off his sorrows, he ordered a dazzling new toy, an electric-powered car, to buzz him about outside. He configured the vehicle so he could lie back with the top down, gazing at the sky above the surrounding rocky peaks as he cruised along. When he realized the battery limited his range, Palmer turned to a diesel-powered Stanley Steamer, essentially a locomotive on rubber tires to go by road. No matter how foul the weather, he was determined to get outside.

One March day in 1909, a snowstorm made a ride impossible. That night, he slipped into a coma and he could not be roused in the morning. He lived only two days more. He was seventy-two. The General was buried in Colorado Springs' Evergreen Cemetery, which he had created like so much else in the town. A year later, his children had the remains of his Queen dug up from where they lay in an English grave-yard, and brought back to be placed beside their father, so she would remain with him in Colorado Springs in death as she never had in life.

⊱⊱⊶⊙⊷⊶⊙⊷⊶⊙⊷⊶⊙⊷⊶⊙⊷⊷

CHAPTER 27

California for a Dollar

DESPITE WHAT HE'D TOLD WHEELER, STRONG ENDED UP using every day of his six-month hiatus, and he emerged from his seclusion a new man. His first letter from the period, dated May 10, 1883, is remarkable just for its handwriting. Where his penmanship had once been a quivery jiggle that drooped wearily at the end of each line, now every letter of every word was so lovingly formed it was nearly caressed, with jaunty flourishes atop his *b*'s and *d*'s. The sentences ran straight across the page unfailingly. His first words conveyed his buoyant mood. "Nothing new here," he wrote. "Weather just becoming comfortably warm."

Newly relaxed, freshly optimistic, he surveyed the railroad landscape that spring—and did absolutely nothing. For once in his life, he decided to let life unfold as it would, and simply wait to see what developed.

In the intervening six months, the Santa Fe's position had not improved. If anything, it had worsened. Even in uneasy alliance with Huntington and Gould of the Frisco, the Santa Fe had been able to steer the A&P across Arizona to Needles, but now that the line was complete, it did little business in the state. Bypassing the Grand Canyon just to the north, the route offered tourists nothing to see, and, stopping at Needles, gave them nowhere to go. If the route had not come with those original land subsidies from 1866, it would have been an utter disaster.

But the newly reinvigorated Strong was fine with all of it. He could see that Huntington was in far worse shape than he was. Strong might have been trying to get into California from one state over to complete his transcontinental route, but his rival had to get all the way to the Atlantic coast. The other three Associates of the Big Four had largely abandoned him, unable to reach consensus on the grand plan. "Two more different men than Leland Stanford and Collis Huntington I never knew," Crocker wrote in exasperation.

Adding to the veneer, Huntington's personal life was a wreck. After his long-suffering wife died in the fall of 1883, he had married his long-time mistress, the brown-eyed Belle Worsham, whose previous husband had run a gambling parlor in Virginia. At thirty-two, Belle was half Huntington's age. For years, he'd been accumulating houses for her off New York's Fifth Avenue at West Fifty-Fourth. Now, as his wife, she required finer quarters at Fifty-Seventh Street, plus a grand country retreat in Westchester County overlooking Long Island Sound. Then she insisted Huntington turn the new city house into one of the grandest mansions in all of New York, adding a central staircase sixteen feet wide, done in jet-black Mexican onyx with rainbow-colored ribbons among other modest touches. As he strained to extend his railroad empire across the country, an economic downturn had begun, greatly adding to his burdens.

Strong let Huntington's problems mount, and then, in August 1884, came to him with a proposition. His Santa Fe was still stuck at Needles, staring across the Colorado River at the Southern Pacific tracks that Huntington refused to let him use. Now Strong decided that enough was his enough. With fresh energy, and renewed determination, Strong reared up and told Huntington that he'd buy the Southern Pacific tracks into California—and if Huntington didn't sell, he'd run in his own line, doubling the SP tracks into California and ruin him. Once, Huntington would have laughed, but the challenges in front of him were too great. He agreed to sell, but only the tracks as far as Mojave, not into Los Angeles itself. He would retain his exclusive on the SP tracks on up to San Francisco from there. To get into L.A., Strong would have to turn to the California Southern, but it was still stalled in San Bernardino, bankrupt and ruined by that canyon washout, eighty miles from any connection to the original Southern Pacific tracks that Strong had just bought.

It would be expensive to get the California Southern up and running, but there was no holding Strong back now. He bought up the line for the Santa Fe, paid handsomely to replace the ruined track, and pushed on into the north to reach the old Southern Pacific line that was now his. The build from San Bernardino required a hard climb over the four-thousand-foot Cajon Pass, and the construction took a full year, but when it was done it gave the Santa Fe a continuous line that ran almost two thousand miles from the Missouri River to San Diego. To celebrate his accomplishment, the Santa Fe named the new town at the junction Barstow in Strong's honor.

Still, it did not give Strong a line into Los Angeles. For that, he had to acquire Huntington's tracks into the city. They departed from Colton, the very place where Huntington had fought so furiously to keep the California Southern from crossing by frog. But by now Huntington saw no point holding them any longer. It was all too clear that Strong was not to be denied, and so he conceded once again.

As a condition of the sale, however, he included the requirement that Strong join with him in a pool to divvy up profits from all the California traffic. In theory, this avoided a price war that might ruin them both, but it ended up putting Strong in a bind, for it imposed on him high ticket prices that kept him from fully capitalizing on his hard-won route to L.A.

To free himself, Strong had to find yet another way into Los Angeles that would have no such constraints. He found it in a local line, the Los Angeles and San Gabriel Valley Railroad, which was building out from Los Angeles to the San Gabriel Mountains past Pasadena, eleven miles due north. It was intended to connect to San Francisco via the original Southern Pacific line past Mojave, but Huntington had been stymying it. For the construction, the LA & SGV had needed to bring in materials on the portion of the Southern Pacific tracks Huntington still controlled, but, determined to keep L.A. an island, Huntington had refused.

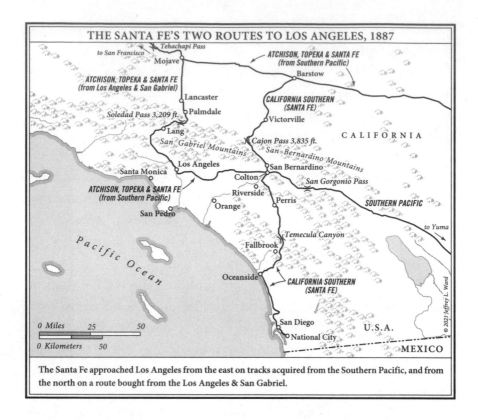

THE SANTA FE'S TWO ROUTES TO LOS ANGELES, 1887

The Santa Fe approached Los Angeles from the east on tracks acquired from the Southern Pacific, and from the north on a route bought from the Los Angeles & San Gabriel.

Strong was only too happy to help. He offered to bring in the building materials from the East along the portion of the old Southern Pacific he now owned—on one condition: that he be allowed to buy it when the work was done. At long last, Strong had his unencumbered route to L.A.

On May 31, 1887, the first Santa Fe train rolled into Los Angeles on these new tracks, breaking open the pool and giving Strong his freedom to make of the city what he would. Done with cooperating with Huntington, now he'd compete. There had been ferocious rate wars before, like Gould's assault on Vanderbilt when he drove the price of a head of livestock on a Central Railroad boxcar down to a penny, but never before had there been a rate war as vicious, or as consequential, as Strong's with Huntington for L.A. Strong went after his nemesis in a blind fury, lashing and lashing him, as if to punish him for all the misery he had imposed. When Huntington charged $125 for a ticket from Chicago to Los Angeles, Strong slashed it to $50. When Huntington matched that, Strong slashed it again. Down and down Strong drove the price, to $8, to $6, and finally just to $1, and Huntington was forced to do the same. If there was rage in it for Strong, there was something else, too—a glorious exultation to have triumphed at last. Strong had brought Huntington low, and he had opened wide the gates to the Golden State. California for a dollar! Such a price was unsustainable, and it didn't last. It drifted back up, cresting at $40. But Strong had made his point. California for a dollar—the price was obviously a stunner, but it was even more thrilling that this was *California* people were getting.

Always before, the Santa Fe's arrival in a new town set off metaphorical fireworks. But at Pueblo, Raton, and the many other towns along the Santa Fe line, the display had amounted only to a bang and a shower of sparks. L.A. was the ultimate, and the Santa Fe's arrival there produced a grand finale of thunderous booms and sizzling meteors and bursting flower blossoms and dazzling curlicues and startling zigzags that lit up the sky not just for the spectators gazing up from below but for the whole country watching from afar. The trains unleashed a torrent of newcomers like nothing America had ever seen, or ever would see again. Four jam-packed Santa Fe trains a day pulled into its spanking new L.A. station, and, not to be outdone, the Southern Pacific sent in no fewer. Between them, the two lines brought in 300,000 people just over the first six months, ten times the city's resident population. The new arrivals filled hotels and boardinghouses as fast as they could be put up, some of the guests reportedly sleeping in bathtubs. And plenty of these newcomers built houses and stayed. Two thousand real estate agents saw to that. By 1890, the L.A. population had shot up to over 150,000, more than five times what it had been five years before, with most of the growth coming since the Santa Fe's arrival in 1887. It made for the biggest surge in population of any city in the history of the United States.

CHAPTER 28

>+⋅>⋅0⋅⊲⋅⊙⋅⊳⋅0⋅⊲⋅⊙⋅⊳⋅0⋅⊲⋅<

Boom!

UNLIKE GOLD, REAL ESTATE USUALLY REQUIRES SOME marketing if it is to command a good price. In the more distant outback, promoters had to work hard to make the land enticing. "What do you want of that vast and worthless area," Daniel Webster had once asked of Wyoming when the idea of a transcontinental railroad was first floated in 1843, "that region of savages and wild beasts, of deserts, of shifting sands and whirling winds, dust, of cactus and prairie dogs?" Since so few people actually knew firsthand what was out there, promoters were free to portray the land as whatever they thought would sell, and there was no limit to some of the claims. To promoters, the West was no Great American Desert. It was, said one, "nature's great flower garden where Eden might have been."

At one point or another, almost every stretch of land past the Mississippi was pitched as a paradise, which led to acute disappointment when newcomers arrived to find nothing but endless winters, unyielding cropland, and social isolation. This had been the cause of Queen's agony. But, to the promoters, reality was no obstacle. If some piece of the West was not actually nature's great flower garden, just wait. It would be soon! The garden would bloom with a little tending. Who could say what the West could *not* become?

Of all the places in the West, Los Angeles was least likely to disappoint. That was its appeal. It was not paradise, but by eastern standards, it came damn close. It had a superb climate—not too hot, not

too cold, but just right practically all year round. A certain P. C. Remondino, MD, took it upon himself to make a formal assessment of
the "climatic, physical and meteorological conditions" of Southern
California, a job, he asserted, he was well-qualified for as a member of
the American Medical Association, the American Public Health Association, the San Diego Medical Society, the California State Board of
Health, and several other august medical organizations. In a lengthy report, he could hardly contain his enthusiasm for the region in which he
now happily resided, and went so far as to say that, of any place in the
world, it offered its inhabitants the best chance for eternal life. Southern California, he said, was a place where "disease and death may be
kept at bay." With its steady sunshine, dry air, blue skies, and delightful
temperatures, it offered what amounted to perpetual springtime. It had
none of the hurricanes, tornados, snowstorms, or lightning that bedev-
iled other parts of the country. In every measurable aspect, the weather
was just about perfect. But it was the sky that got him:

> The sky is of a clear, bright blue, and at times a whole series of
> months may pass without a speck of cloud to mar its surface; it is this
> bright sunshine and perpetual clear, blue sky that is the real uncon-
> scious power—more so than any local custom or habits—that brings
> Californians back to these shores, who, after many years spent on the
> coast, have attempted to again live in the East.

Such boundless enthusiasm for California was not new. A decade
before, the Southern Pacific hired a newspaperman named Charles
Nordhoff to extol the entire state to easterners in *California for Health,
Pleasure, and Residence*. For ease, comfort, and beauty, he claimed,
California was a perfect extension of the state-of-the-art Pullman Palace Car that delivered them. The state was completely enjoyable—
beautiful, clean, relaxed, everything a person could want for a better
life. Given his employers interests in San Francisco, Nordhoff played
up the attractions of the northern half of the state, and largely skipped
over the south, saying only that it was for now best suited to "invalids."

Now, with trains full of newcomers pulling into the city, a Los Angeles journalist, Benjamin Truman, was charged with drumming up tourist business for Southern California for a promotional book. He deemed the region "semi-tropical," but about halfway through his book it seems to have troubled him that while "semi-tropical" might have pleasing associations of a balmy Pacific island like Tahiti it might betray a hint of sloth, perhaps even of decadence, that might not appeal to the emigrant Puritans from the East who were his primary audience. So Truman shifted his reference point to the more cultured Mediterranean. Spain, the south of France, Italy, Greece—all of them offered a reassuring impression of settled, sunny permanence, the terraced orchards evoking an equally cultivated citizenry.

It was left to a later journalist, Charles Dudley Warner, to call Southern California "our Italy." The designation brought to mind such cultural hotspots as Venice, Florence, and Rome, evoking what a later writer has fancily termed the "Mediterranean littoral" of olive groves by the sea. It reached back in time through the Renaissance to the glory days of the Roman Republic and could be easily visualized, since by then countless paintings of the Italian countryside had gone up in the art museums of the citified East. Warner argued *our* Italy was even better than the original for *our* Italy was "without marshes and without malaria," and *our* Italy would be productive, its soil to be worked, and not let to go to seed as the Italians seemed to have done. Thrilled by this notion, Warner fairly shouted it in his text: "Here is our Mediterranean! Here is our Italy!"

⊱┄◦┄⊰

The grand vision took few years to fully settle in. Initially, the frenzy for Los Angeles real estate, sparked by the miracle of California for a dollar, was oddly formless but was such an electrifying phenomenon that it acquired a new word to describe the frantic buying: "boom!" (usually with the exclamation mark included). There had been real estate bubbles before, but they had always popped. L.A. real estate, and the land around it, really was worth buying at ever-higher prices—and, indeed, they've almost never come down since.

The boom had its publicists in town—every real estate salesman and developer doubled as one—but the unusual thing was that it had infinitely more boosters all over the country. It seemed an entire industry had sprouted up to promote the wonders of L.A. in printed matter of every type—brochures, posters, features, editorials, newspaper items, all adorned with copious illustrations of the good life and detailed maps showing potential real estate buyers what was where. Of all the endorsements, though, by far the most effective were the letters back home from people who actually had moved to L.A. They were so delighted with their new lives in the warm air, they wanted their friends and family to join them.

In just the first six months of 1887, a staggering $100 million worth of Los Angeles property was sold. A typical lot on Seventh Street in downtown L.A. zoomed from $11,000 in 1886 to $80,000 a year later,

Los Angeles in 1891, after the Santa Fe arrived.

post Santa Fe. The venerable pueblo turned itself into a true city almost overnight, as plans almost immediately came forth for a new city hall, a new courthouse, more schools, proper sewers, and, finally, paved streets.

The exuberance had its lunacy, to be sure. There was supposed to be a brand-new Border City on the edge of the Mojave Desert to be reached by hot air balloon, and a Gladstone, named after the illustrious British prime minister who was claimed, erroneously, to have bought in. One contemporary historian of the boom, T. S. Van Dyke, satirized such hoopla by evoking a bogus new town he called Balderdash on the side of Mt. Baldy in the San Gabriel Mountains overlooking L.A. He could well imagine the brochure's headline, for there were many like it:

BOOM! BOOM! BOOM!
The newest town is out! Balderdash! Watch for it! Wait for it! Catch on to it!

But even a skeptic like Van Dyke had to concede that Los Angeles had a lot to offer and couldn't help waxing lyrical as he went about the view from a fictional Excelsior Heights.

A medley of colors was blazing over hill and dale, acres of poppies with lustrous orange tints, acres of golden violets whose fragrance filled the breeze, everywhere the delicate pink of the alfileria and the tender blue of innumerable bell-flowers, with white and scarlet and purpose rolling in gay confusion over the plains and up the feet of the hill to where the crimson of the wild peak, the lavender of the lilac, and the carmine of the wild gooseberry lit up the dark green of the chapparal.

There was no getting around it. L.A. really was pretty nice.

Between January of 1887 and July of 1889, sixty brand new towns came into existence in Los Angeles County, twenty-five of them along the Santa Fe tracks to San Bernardino. They appeared "like scenes conjured up by Aladdin's lamp," went one contemporary account. They popped up everywhere—"Out of the desert, in the river wash, or a mud flat, upon a barren slope or hillside." It seemed the Santa Fe created a land boom wherever it went, creating handsome, thriving places like Lincoln Park, Monrovia, Glendora, Altadena, Duarte, and Pomona, whose Congregational Church sprouted a college that then spawned Claremont and four more. In his excitement, Strong sent tracks nearly everywhere in greater L.A. He ran a line out to the Pacific coast to build up Santa Monica, turning the site of the early American colony into a hotspot, and another southwest to Redondo to inspire a spectacular hotel on the beach. He sent yet another southeast to Santa Ana and then farther down the coast to San Diego to give that city a second train, along the way building up Anaheim, previously just a vineyard tended by a few hundred German immigrants, the Quaker-founded Whittier, and the new city of Orange. He even sent a train out just to do a crazy loop around newly burgeoning Riverside.

Pasadena's "Royal Raymond" hotel.

In the general makeover of greater Los Angeles, nowhere was more illustrative of progress than Pasadena. Exactly as the residents hoped, the Santa Fe had put that town on the map. The first grand hotel, the Raymond, or "Royal Raymond," went up on its sole hilltop in anticipation of the train's arrival and quickly became *the* place for dances and concerts. It was named for its builder, Walter Raymond of the early Boston travel firm Raymond and Whitcomb, that sent in tourists. Raymond had persuaded Strong to put the Santa Fe train station at the foot of the hill for easy entry up to the hotel, and the location still left room for another swanky hotel, the Green, to push in closer to the Santa Fe to nip off passengers for its new, high-towered, Romanesque establishment. A third, the Painter, swooped in to scoop up the overflow from the previous two soon after.

Together, the three hotels created such a buzz that when the city put some land by the Raymond up for sale, investors engaged in panic buying, egged on, perhaps, by the fifty-one realty offices that had opened up in town. One lot that had sold for $170 in 1886 went for $10,000 two years later. A man who'd bought ten acres by the Raymond for $550 in 1875 sold one and a half of them for $36,000 in 1887. Five trains a day ran between Pasadena and Los Angeles, the influx turning a village of a few hundred horticulturalists into a city of six thousand in just two years.

Imported easterners, eager for sun, started trying to outdo each other in the manner of the plutocrats on Fifth Avenue as they turned simple bungalows into the Greene and Greene beauties of the early Arts and Crafts movement, and then into Victorian mansions complete with gables, stained glass, porches, and exotic stone facades. And the city dressed up accordingly, as once-dusty ranchland evolved into a garden city that took the rose as its emblem, soon parading it before the nation every New Year's Day, an annual teasing reminder to shivering northerners of Pasadena's eternal warmth.

>-+-•--0--•-+-<

Despite its reputation and appeal, Pasadena was far from the only garden in Southern California. If promoters were always trying to pass off the western wilderness as a garden in the making, in Southern California they didn't have to do much pretending. While the region, once it was irrigated, had been known for its soil, it had never been known for the *fruits* of its soil, they were so uneven in quality—until one fruit emerged from the fray, a perfectly round, succulent orb, named for its color. If Eden had the apple, Southern California had the orange. And it was Strong who saw to it that the entire United States should get a chance to eat one. The orange would be greater L.A.'s tasty, succulent ambassador, borne back to the East in the refrigerated cars of the Santa Fe to easterners eager for something sparkling on the breakfast table.

Florida had produced its own oranges, of course, but its groves were periodically wiped out by frosts, and they had none of the delectability of the California version. The first crop was grown in Riverside, an agricultural colony founded in 1870 with the expectation that a second transcontinental railroad would soon come through. In the meantime, the Riverside founders tried a variety of commercial produce, from walnuts to opium, before settling on oranges.

No one found exactly the right variety until an unlikely farming couple wandered in, the Tibbetses. Eliza, remarkably, was a spiritualist who claimed to be guided by a long-dead American Indian maiden named Floating Feather. Her husband, Luther, had initially come to her to contact his late wife, the sister of his previous one. Shortly after they arrived, they received a Brazilian varietal from a friend of theirs who happened to be the United States Department of Agriculture's chief botanist. He thought the couple might be interested in the Brazilian, an unusual "navel" orange, named for the button-like tip, that was uncommonly juicy and sweet, and also hardy because of its thick peel, and seedless besides. The botanist friend thought it might be just the thing for a climate like Riverside's. Eliza Tibbets planted one beside her house, splashed it with dishwater, and awaited its first fruit.

To the Tibbetses' disappointment, the first oranges were lumpy and dry, but the couple persevered. Before long, the Brazilian navel took to

California's sunshine and soil, and put forth some fruit that seemed to capture the spirit of the place.

After the Tibbets Navel, as the couple took to calling it, won first prize at the Riverside citrus fair in 1879, a local entrepreneur laid out a seventy-five-acre grove of the fruit, and soon it seemed every grower was throwing over their walnuts, almonds, and prunes in favor of the fantastic new orange. They put it up against Florida's best at the 1885 World's Fair in New Orleans—and emerged the global winner. Orange-growing towns sprang up everywhere in what would be called Orange County, and residents organized a national citrus fair in Chicago to show off what was now billed as the Riverside Orange. A British syndicate, the Riverside Orange Company, came to town to add thousands of acres of orange groves and create a vast system of canals to irrigate them. British aristocrats followed, founding the Riverside Polo and Golf Club, a choral society, the Riverside Orchestra, and the Loring Opera House, which opened in 1890 with a boisterous performance of Gilbert and Sullivan's *Iolanthe*. When the town of Riverside in turn gave birth to Riverside County, it created as its courthouse a full-scale replica of the Grand Palace of Fine Arts of the Paris Exposition of 1900, a neoclassical extravaganza dubbed a "Roman Imperial wedding cake." By 1890, Riverside alone shipped out by the Santa Fe 1,253 carloads, or 358,341 boxes of oranges, half the total of the entire state. Twenty years later, the number of boxes was up to 2.3 million.

It was to capitalize on the product that Strong created his giddy one-hundred-mile loop to Riverside about the orange groves of the San Gabriel Valley. He wanted to show off the orchards to tourists and to ship the fruit out on his new refrigerated cars. But the oranges were just the start of California's growing reputation as America's orchard. Soon came lemons, peaches, plums, apricots, pears, nectarines, every sweet thing that could come from the ground, turning greater L.A. into "the garden spot of the earth," as more than one promotion had it.

And, incredibly, it really was. It also produced flowers—colorful, exotic ones like bougainvillea, magnolia, grevillea, hibiscus, and that imported glory, the rose, that inspired a parade, all of which established

L.A. as a paradise of fragrant colors. With the flower beds, in time, came the lawns, lush green carpets that unrolled from every home.

Trees, too—not just the chestnuts, elms, and pines that reminded the easterners of home, but more exotic trees that evoked fantasies of elsewhere. The eucalyptus was extremely popular, but another tree grew up everywhere so thickly, so toweringly, and so strangely that it came to define visual Los Angeles: the palm, particularly the date palm, a shaggy, fat-trunked profusion of fronds that would come to fill every viewfinder and postcard in the city. Hauled in by the thousands as seedlings on railroad flat cars, it was spread just about everywhere, along streets, in parks, beside buildings. Elephants and brass bands might have sold these new developments the trains brought, but the palms gave them their distinct personality. Originally to be found in the hothouses of wealthy easterners (Jay Gould was especially partial to them), they gave every boulevard the aura of a Millionaires' Row.

By the time of Chicago's famous "White City" world exhibition of 1893, that palm so symbolized California that the Santa Fe railroad carefully uprooted a massive descendant, all fifty feet and forty-five thousand pounds of it, gently placed it aboard two flatcars, and hauled it two thousand miles to display it in triumph to the world at the fair. When the Santa Fe dropped off its passengers by the Hotel Green in Pasadena, it sent them through a palm garden to complement the station's otherworldly Moorish architecture. The Santa Fe also went in for a Moorish dome-topped fantasy when it replaced its original 1887 L.A. depot, and bordered it, too, with a stand of palms that told passengers they'd come farther than they'd dared imagine.

CHAPTER 29

Home

AS THE POPULATION OF THE CITY OF LOS ANGELES GREW, so did Strong's Santa Fe railroad. Its line now stretched back to Chicago, and nearly everywhere else, like a hand up to the northern Midwest, down to Texas, and deep into Mexico. And, of course, it connected to innumerable other lines of other companies all over America. At each western stop, the Santa Fe had brought pioneers about as far west as they could go, but Los Angeles, of course, was the ultimate. The Santa Fe trumpeted the railroad's transcontinental achievement with the proud words, *East or West, Santa Fe Is Best* inside its fabulous new LA train terminal erected in 1893. By then, it was receiving fifty-two trains a day from points all over America. Strong also made sure that the Santa Fe got in on the global promotion of the ruby-lipped Nellie Bly when she tried to outdo Jules Verne's *Around the World in Eighty Days* in 1889, riding the Santa Fe for the portion from California to Chicago, ultimately beating Verne's Phileas Fogg (modeled on the Union Pacific's "Mr. Train") by slightly over a week when she pulled into Jersey City to complete the global circuit. That year, the Santa Fe's total track count came to 8,118 miles, enough to take it a third of the way around the world and make it by far the biggest railroad in America.

In overspreading the country as it did, the Santa Fe was finally accomplishing Lincoln's dream of unifying it—not so much by laying down track as by shipping goods everywhere those tracks ran, and thus integrating the United States into a common national market. In this, curiously, the model was time itself. Before the transcontinentals speeded passengers West, time was a local affair, determined by the exact moment that, according to local astronomers, the sun was at the top of the sky. So it was that when it was noon in Boston, it was still 11:47 in New York City, and so on around the country. This was little more than a small inconvenience so long as the country's transportation lines ran north-south with the waterways. It became a major nuisance when it started going east-west with the trains, and travelers had to adjust their watches dozens of times on the way to California. To resolve the matter, in 1883 the trains introduced "Railroad Time," which divided the country into just four "time zones" to accommodate the sun's passage across the sky.

By then, the trains were transforming the national psyche similarly to forge a single identity based on the goods that were now transported to every region of the country. It started with the Santa Fe's Harvey Houses, which went national as the railroad did, bringing a uniform standardization with them. By running everywhere about the country, the Santa Fe created new national distribution channels for merchandise, too, giving any product anywhere a chance to sell everywhere.

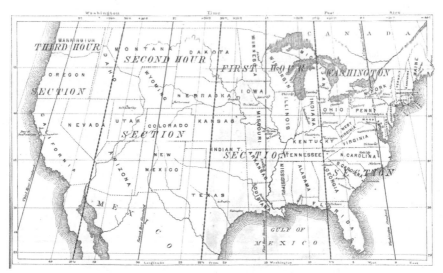

An early map of the railroads' National Time.

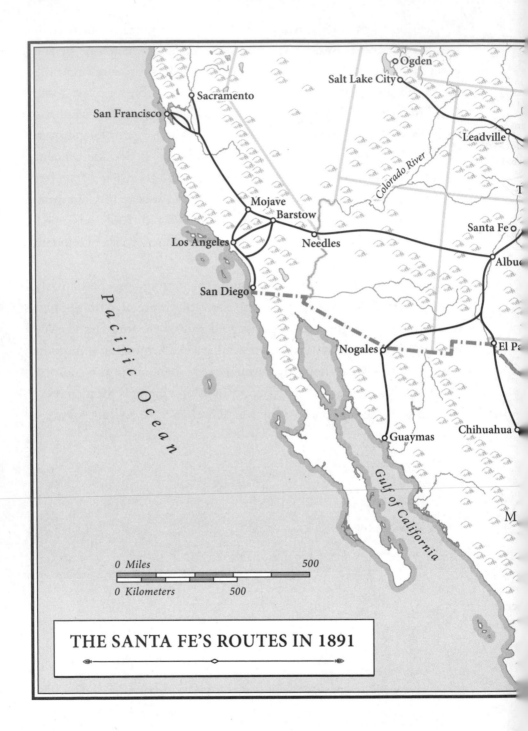

THE SANTA FE'S ROUTES IN 1891

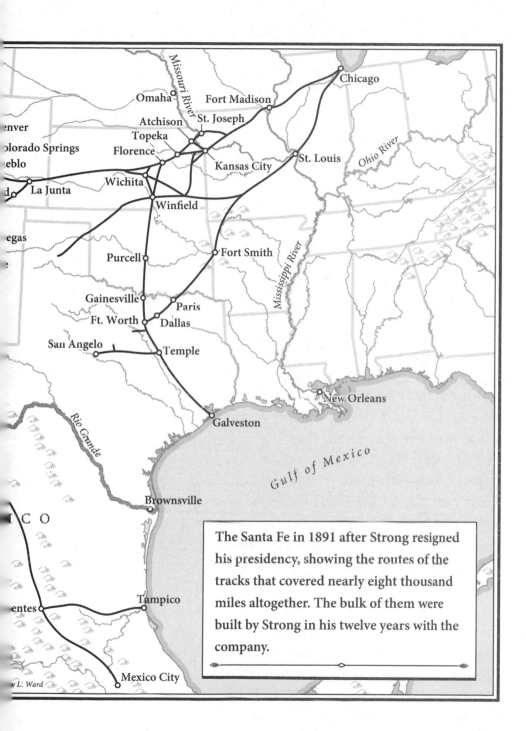

The Santa Fe in 1891 after Strong resigned his presidency, showing the routes of the tracks that covered nearly eight thousand miles altogether. The bulk of them were built by Strong in his twelve years with the company.

Anyone could now buy anything, starting with the offerings of the Montgomery Ward's mail order catalog. Dreamed up by a down-on-his-luck Chicago salesman, it bypassed retailers to sell products directly to customers all over the country, allowing everyone to get in on trends that had previously been confined to big cities of the East, creating a national consumer sensibility based on standard fashions, regular sizes, and common colors. Other products followed. Chicago lumber was distributed throughout the West. Once just a local offering, beer was taken national by Adolphus Busch, cofounder of Anheuser-Busch in St. Louis, in the form of Budweiser. To sell it, he created a national image soon termed a "brand" that sprang free from its local origins. Many more followed—Levi's, Ivory Soap, Coca-Cola, Kellogg's Toasted Corn Flakes . . . Such brands were propounded everywhere in the advertising that filled the newspapers and magazines that the railroads delivered to readers throughout the country.

As the Santa Fe grew, though, so did other lines, eventually overspreading much of the country, and forcing nearly all of them to face that elemental question—cooperate or compete? By then, cooperation meant working together with a number of other lines, not just one, which was difficult enough, because it required the creation of a pool to set common rates like the one that Huntington had with Strong in California. It was hard to create a pool that worked equally well for all parties, and, embittered by his experiences with Huntington, Palmer, and Gould, Strong steered clear. He imagined he could use his size to his advantage to command his price.

But as other lines filled in the West, the Santa Fe was left with fewer exclusives. While Strong had the most track, he also had the most track to pay for. When he added more and heavier freight cars, he had to upgrade all his track from iron to steel to withstand the load. To accommodate higher speeds, he had to straighten out curves, flatten grades, and strengthen bridges. To match rising expectations of comfort, he replaced or renovated all his passenger cars. He had to add employees, too, and raise salaries to avoid strikes and maintain loyalty. It was expensive, and revenues were not keeping up. That proved a problem he could not solve by laying more track. Already squeezed by the competition, the Santa Fe's revenues were further damaged when the Midwest suffered a devastating drought in 1888. Unable to sell stock to cover its losses, Strong had to issue high-interest bonds to stay afloat. The balance sheet turned an alarming red.

An optimist-turned-fatalist-turned-optimist once more, Strong relapsed to pessimism, bracing himself to the inevitable. In his report to shareholders in 1888, he adopted a newly elegiac tone, the sound of farewell.

> *The history of western railroad construction for the past quarter of a century has demonstrated that successful results can only be attained by occupying territory promptly and often in advance of actual business necessity. This was the policy of the Atchison company from the first. It led the way. It built, not upon assured returns of profits, but upon a faith which time has abundantly vindicated—that the great western and southwestern regions of the country were rich in possibilities and the company which first occupied the country would reap the first and greatest rewards.*

He went on:

> *There is no wisdom so common as that which comes after the fact. Droughts, failure of crops, excessive competition, continually decreasing rates, unwise legislation, strikes and other calamities have befallen us as they have other western roads. But your directors could not know in advance that any of these unfavorable conditions would have to be met, much less that they would all have to be met at one and the same time.*

Virtually every line in the Santa Fe system was losing money, some of the losses running into the millions, and putting the company in

serious debt. Santa Fe stock plunged from a high of just under one hundred to a mere twenty-six and three-eighths. For the first time, industry
analysts turned against Strong. *Railway Age* called him the Santa Fe's
"Napoleonic president." A prominent investor declared the company
"an utter wreck," and the *Railway Review* derided Strong's expansion-
ism as a "mania."

If Strong had once drawn strength from his company's resources as
a broad-based corporation, he now found that a point of vulnerability.
The investors from the Boston Crowd who'd first hired him had owned
a majority of the company, but now their holdings were relatively
scant. In the vast enterprise that is a public company, their influence
was meager. The Santa Fe relied on Kidder, Peabody to raise capital,
and awarded it seats on the board to reflect its importance to the railroad, but now the investment house was afraid the Santa Fe's mounting
debt might drag it under. In January 1889, Boston stockholders started
to turn against their railroad, flooding the Boston Stock Exchange—
where Santa Fe stock accounted for 70 percent of all transactions—
with sell orders. They pressured the company to remove from the board
the five members tightest with Strong and replace them with men loyal
to Kidder, Peabody.

That was it for Strong. Refusing to be reduced, as he put it, to an
"errand boy" of Kidder, Peabody, he tendered his resignation on September 6, 1889, and this time the board accepted it. He'd reached the
end of the line.

The official history of Strong's native Rock County, Wisconsin, attributes its favorite son's departure from the Santa Fe to "failing health"
not to any forced resignation, so he may have succumbed to "the blues"
once more, but it's hard to know. Strong and his wife returned to the
Beloit of their former days and bought a small farm he called The Par-
tridge after his wife's family on the outskirts of town. Strong briefly took
over the presidency of a local bank and created a subdivision called

"Strong's Addition," where he gave names from his own family to four of its streets, one of them Strong Avenue. He also expanded the town cemetery, reserving a plot for himself and his family.

It's quite possible that Strong had accepted his departure from the Santa Fe after nearly twelve years of loyal service and felt no bitterness on leaving the railroad he'd built. He'd taken what had amounted to a local line that barely made it out of Kansas and defeated not just Palmer but Huntington and even Gould to run his railroad all the way to the Pacific, exactly as he had long intended. In the process, he created one of the largest railroads in the world, and some of the major cities of the west, culminating in Los Angeles, which he turned from an obscure pueblo into the second-largest city in the United States.

As ever, his actions best express his true feelings. In 1908, he and his wife left Beloit and moved to Los Angeles themselves, where they spent the remainder of their days. To get there, the president of the Santa Fe provided them his private railroad car, something Strong had never possessed.

The couple moved into a modest apartment on South Union Avenue near the present MacArthur Park, just west of the downtown that was once the heart of the original Spanish pueblo. They took rooms in two other apartment buildings before settling into a modest bungalow "away from the noise and confusion of street-car traffic," a few blocks northwest in a place now called Rampart Village, that had been developed by a wealthy couple who imagined themselves illustrious for owning much of Catalina Island.

By then, Strong's wife had died, and his eyesight was finally failing him. He required the day's railway news to be read to him by one of his several attendants. An occasional stroll or visit from his children and grandchildren were his only other diversions. Still, even as his sight dimmed, his friend recalled that Strong had a way of looking right into a person with his "wonderful eyes," while a "delicately sensitive expression" played at the corners of his mouth, the encounter to end invariably with a "forceful handshake" that suggested a "gratuitous humanity."

The end came on August 3, 1914, when Strong was seventy-seven years old. Prepared as ever, he'd made sure that all his friends be given good seats for his funeral service in the Congregational Church in Beloit "irrespective of wealth" and, as a God-fearing man, he preferred that his final train journey north not be conducted on the Sabbath.

The Frontier Thesis

AMONG THE MANY TOWNS STRONG'S SANTA FE BOOM created in Los Angeles County, one of the most notable had been an orchard growing figs and apricots in the Santa Monica foothills. It was selected for development by a Topeka, Kansas, couple, Harvey and Daeida Wilcox, because they liked the light. They named it Hollywood for some holly bushes by the house. They'd intended to create a good Christian liquor-free colony of the sort that had once been common in the region, and the site attracted an unusual number of churches—Southern Methodist, Episcopal, and Methodist. But it remained largely cut off from the city of Los Angeles until a rough road called Sunset Boulevard was laid out in 1904.

Hollywood might have been lost in the blur of Los Angeles County if David Horsley hadn't come to town. Horsley was in the nascent movie business, working with a refugee from Thomas Edison's latest venture, the Biograph Motion Picture Company, to form the Centaur Film Company, to crank out jerky one-reelers on western themes. The first one was called A Cowboy Escapade, and it set the template for all the others. His company produced one film a week through the summer until the fall, when the light started to fade. Indoor lighting was not yet developed for filming, and in the autumn of 1911, Horsley decided to follow the sunshine to the Wilcoxes' Hollywood, where he set up shop at the corner of Sunset Boulevard and Gower Street. He continued to make Westerns, starting with The Law of the Range. He taglined each

Hollywood in 1887.

one "A Western from the West," but the motion pictures became better known as "Horse operas," tales of cowboys without Indians, all of them shot in the Hollywood hills, which could pass for the American West if you'd never been there. (It's hard to imagine that the term didn't lead to a few puns at Horsley's expense.)

The Universal Film Company bought out Horsley the next spring, making it part of the studio that became known as Universal Pictures. By then, such luminaries as D. W. Griffith and Cecil B. DeMille had descended on Hollywood as well. They came by train, as did all the directors, stars, and backers, bringing in the aspirations that rose out of the vastness of America. They deposited those desires in Hollywood to distribute to silver screens everywhere.

As such, the movies were the natural culmination of the railroads, the Santa Fe not the least of them, starting perhaps from their earliest days, when passengers gazed out the windows, in pursuit of their dreams, idly marveling at the still panoramas in the distance, as things closer by whipped past. Those images might have been the first movies of the West, with the railroads the things that moved. But the movies drew on years of the highly fanciful promotions that turned California, and particularly Southern California, into a place of the imagination for people who seek something more, whatever it might be. A place of dreams for dreamers.

>─┼─◆>─○─<◆┼─<

In 1893, a young, little-known historian named Frederick Jackson Turner at the University of Wisconsin produced a lengthy paper at the annual meeting of the American Historical Association in Chicago that stemmed from a statistical item from the census report of three years before. It indicated that the American population was now sufficiently dispersed across the country that there was no longer a "frontier line" with any significant open space beyond it. The frontier had vanished. The West was now closed.

To Turner, this changed everything, for he believed that America drew its distinctive qualities from its western frontier—first by having one, and then by conquering it. He saw it as the inexorable advance of civilization over the wilderness, but one in which the wilderness transformed the civilization that was subduing it. It forged the unique American character, one drawn from the ruggedness, practicality, pluck, ingenuity, and occasional violence needed to survive on the harrowing frontier. This is what made Americans so different, and, to Turner, so much better than the people of other lands.

This became known as the Frontier Thesis, and for a long time it was for American historians what Darwin's theory of evolution was to biologists. It remade the popular understanding, and even became a point of national pride. Of late, though, that enthusiasm has faded as critics have questioned this self-congratulatory creation myth by which our forebears became heroes of the prairie for conquering the elements and slaughtering the "savages" who'd come before.

But there is another curious omission in the Frontier Thesis. In his lengthy disquisition on the settling of the West, Turner makes only passing reference to the railroads, even though he must have realized that the frontier closed about the time that the railroads had filled every corner of the country with their tracks. (It's no coincidence that Strong had been forced out of the Santa Fe just the year before, when his railroad had reached its limit.) And Turner nowhere mentions the innumerable towns the railroads created, fully platted and ready for these intrepid pioneers to inhabit, to say nothing of the ersatz government they brought to the West before it was formed into proper states. This is part of what is termed the "industrialization" critique of the thesis. Turner does not much delve into it because it contradicts his belief that the West was won by men, and not by machines. But, of course, it was won by both, men using machines, even as—Marx would not be the only one to point this out—the machines were using them.

Both sides of the argument overlook another core element that doesn't fit with either position: the power of the driving dream behind this relentless charge to the West. The original settlers, mostly from Europe, came west across the Atlantic to find something, and, once they settled on the East Coast, their descendants were tempted to search farther west still, and, as Huck Finn put it, "light out for the territory," for escape and betterment, both. On and on they went until they hit the Pacific. These pioneers were inevitably affected by the struggle to achieve their desires, but they were also driven by the conviction that, unlike other countries, American had a paradise to be found somewhere beyond the next hill.

That was the conviction that powered Strong and the General, too. If nothing else, they were both restless, questing men, never fully satisfied that they had gone far enough. Every promotion of their railroads, and all railroads, culminating in the most spectacular ones for California, fed this belief in a better tomorrow elsewhere. To a great extent, this represented the distinctively American pursuit of happiness.

What is so striking about the United States is that after the West was won, those dreams did not die. They merely relocated. Once Americans reached the Pacific, they put them up on movie screens everywhere. It wasn't just the light that the early moviemakers needed from the West but the free play that at last land gave to their imaginations. Flickering in the dark, only seeming to be real, a movie is perfect for fantasy. And those fantasies transfixed Americans no less than the prospect of frontier adventures did their ancestors. It was a shallower engagement, to be sure, but it offered something similar, about lives not yet lived, of places not yet visited, of feelings not yet felt. That was always the soul of the West. It was the far shore of the imagination.

►─◄─○─◄►─◄

Sadly, the trains that took the country west, like passenger trains everywhere, exist today mostly in memory. After being taken over by Amtrak in 1971 with the collapse of the Penn Central, too many of the lines that remain have all the charm of the post office. While they remain a pleasure for their riders, they've been viewed by others as examples of federal overreach, their budgets trimmed, their routes limited, and the thrill dimmed. For many, the glorious passenger trains of a former day, the Santa Fe and Rio Grande among them, have all been reduced to objects of nostalgia buzzing about atop basement ping-pong tables for the amusement of children. Or they are revered for the exactness of their engineering, as hobbyists savor the exquisite differences between the 4-4-0 and the 4-4-2 configurations of wheels on the early locomotives, or the precise configuration of railroad stations in 1870s Colorado. The one place where railroads are most widely valued, curiously, is on the Monopoly board, where four historic railroads govern the four sides, elements of life as central to existence as property or ill fortune, but, alas, ridden only by Chance and useful mostly for their contribution to the revenue stream, especially in combination.

But if the railroads themselves have largely disappeared from the landscape, they have left notable descendants in the "platform firms" of Amazon, Google, Facebook, and others that have transformed all life as the railroads did before them, the electronic platforms the new tracks. If the railroads closed distance, sped up time, conquered territory, seized markets, twisted the national economy, repurposed the workforce, and altered mindsets—one would say no less about their Internet counterparts. And, like the railroads before them, many of these platform firms relied on government contributions (Apple's free

use of the Defense Department's mouse-based interface, for example)
to grow and then dominate their "categories," be they search or retail,
that they became governments of their own, answerable only to themselves. Although the number of dominant tech firms is comparable to
the national railroad lines, the Santa Fe among them, their might is
exponentially greater because their reach is global and their hold over
their market so complete. As a result, the Big Tech billionaires possess fortunes infinitely greater than the richest railroad tycoons. Cornelius Vanderbilt's then-staggering $100 million would today be the
equivalent of $2.5 billion, no small sum, but a scant fraction of Jeff
Bezos's personal fortune, at last count nearly $200 billion. Such sums
are easily converted to political power that can hold governments at
bay. For the railroads, the concentration of wealth and power inspired
a political shift from the laissez-faire policies of the Gilded Age to the
trust-busting of the Progressive Era as government responded with reg-
ulations to curb the most egregious abuses, and a temporary takeover by
the Interstate Commerce Commission. Done in the name of the "public interest," the regulatory process revealed there was no such thing,
only a collection of special interests that competed for "regulatory capture" by which their own interests were favored over all the others. The
results proved dangerously restrictive to the railroads' ability to adapt
to a changing market. When the country needed to mobilize for the
first World War, the railroads were legally unable to comply. And then
cars, buses, and trucks came along to do to trains what trains had done
for canal boats, while riding on roads created largely at public expense.

Most strikingly, the platform firms have transformed reality even
more drastically in our time than the railroads did in theirs. Nothing
could have been more real than a locomotive. It was not just seen, but
heard, smelled, and even felt in the rush of air as it blew by. Not so with
the digital version, riding imperceptibly from anywhere to everywhere,
drawing to their on-screen products seemingly everyone's eyeballs—
and, of course, their brains, too, altering the fundamental nature of
experience as reality turns virtual.

There is a kinder point of comparison, though. Nothing so popular is all bad. Strangely seductive, the messages warming our screens hold out the promise of something better, if only a speck of good news. We turn to our cell phones in hope and for escape. In the entrancement of that glowing screen, created for the most part in California, it is not hard to see a vestige of that original dream that Americans chased into the West. If that was a vision of paradise, the imagery on cell phones, tablets, and computer screens is a glittering virtual world that rivals, if it does not actually eclipse, the one we live in. But it shows how much smaller our lives have become now that the frontier is closed and the vivid promise of the West has yielded to tiny but brilliant images we can hold in our palms.*

However sweeping the western railroads may be in their ultimate implications for capitalism, technology, marketing, politics, globalism, and so much more, the story of the Santa Fe and the Rio Grande is still largely the story of their leaders, Strong and the General, and it took the course it did largely because of them. Like the proverbial hedgehog, Strong was a man of one idea, Grow or Die; like the proverbial fox, Palmer had possibly too many ideas, that extended from his romantic notions of the West and his joy in plunging "into the maelstrom," to his love of the local and his desires for his Queen.

Palmer's quest for glory ended sooner than Strong's, with less to show for it, while Strong achieved about everything he sought, but both came to realize that there were limits to the rules they lived by. Strong's belief in big made his line too big. Palmer's faith in small kept his line too small. In some ways, the lesson in these two men's fates is as simple as that. But the more complex truth lies in the strange

---

* Writing this in COVID times, I should add that the railroads also did their bit to create the present global pandemic, if only inadvertently, for they contributed toward the globalization that has now exposed almost everyone on earth to the peril of the novel virus. In this, too, the platform firms have taken over from the railroads to market their wares not just across the continent, but around the world. Their electronic pathways carry only data, of course. But where data goes people follow, and vice versa.

intermingling of these two men's souls. As each tried to seize what the
other wished to claim, they revealed a secret bond. It was almost as if
these two ferocious antagonists were actually the same at heart: They
were roused to war not by their differences, but by their similarity—one
shared by the ambitious everywhere in their insatiable desire for more.

▸┼◆─○─◈┼◂

Today, the Rio Grande has been reduced to a tourist line that chugs
along narrow-gauge tracks up the Royal Gorge but stops well short of
the twenty-mile Dead-line and doesn't come anywhere close to Leadville. But it does pass over a remarkable "hanging bridge" that suspends
the tracks over the Arkansas River, and offers a view of the stone forts
that crews from the Santa Fe and Rio Grande put up on either side of
the gorge to scare each other off.

The Santa Fe carried on for decades more, with a temporary resur-
gence that pushed its tracks to more than thirteen thousand miles. The
story of the Santa Fe has been brought to the movies as *Santa Fe* and
Denver and Rio Grande, about its fight for the gorge, not to mention
Judy Garland's spicy *Harvey Girls*. But, after Strong left, it never came
close to recapturing the thrill of charging west or the glory of bursting
into Los Angeles. It dwindled in the new century and after a series of
increasingly desperate mergers, some with portions of its old nemesis,
the Southern Pacific, it finally became nothing but a memory. Amtrak
took all the Santa Fe passenger lines in the nationalization of 1971, and
the Burlington Northern—itself the combination of the old Chicago,
Burlington & Quincy and the Northern Railway—claimed the last of
its freight lines in 1995. Although the new company graciously tacked
on the Santa Fe's initials to call itself the BNSF, the Santa Fe is no
more.

General Palmer's spirit lives on in Colorado Springs and other
places in the Rockies, where history museum gift shops offer his image
emblazoned on coffee mugs and dishtowels. Although he was far more
accomplished, Strong has faded almost entirely from memory, his name
easily recalled only at Barstow, California. Still, his legacy is every-

where. Not only in star-studded, ultracool Los Angeles, but through-out Southern California and across Arizona, New Mexico, Colorado, and Kansas, and into Mexico, too, along the routes he plotted that are recorded in this book, and many others down to Texas and into the Midwest. In the course of laying more than eight thousand miles of track, Strong built or acquired so many subsidiary companies it took the railroad's corporate history, *The Story of the Santa Fe*, forty pages to list them. Not all of these were Strong's doing, but plenty were.

All lives touch other lives, and both of theirs, Strong's and the General's, touched far more than most. Their distinctive, rivalrous spirits, backed by their industry, persistence, and daring, defined the West and remade a good deal of the nation, too. Both cautionary and inspiring, their entwined lives come down through history to the present moment where the great American experiment is poised as always on the brink of the future. And they should inspire us to be bold, lay fresh tracks, and reach for a better life ahead.

Acknowledgments

The history of the Santa Fe and Rio Grande Railroads is like their tracks—still in existence, just hard to find. Some of the original trunk lines, renovated to modern standards, still carry freight, but precious few take Amtrak passenger cars, and none of the branch lines into the mountains have been used for decades. As I traveled through Kansas, Colorado, and New Mexico, following the routes of the two lines, I was always thrilled to spot the original rails. They were invariably rusty, the ties splintered and overgrown with weeds, but seeing them I imagined I'd come upon a historic battlefield like Valley Forge or Gettysburg.

I had a grand time driving where the Santa Fe and Rio Grande went, and, if any readers are curious to see the territory Palmer and Strong fought over, they should do the same. It was particularly great to bomb down Interstate 25 in a rental car from Denver to Santa Fe, hitting the hot spots of my narrative, and I'm grateful—as everyone should be grateful—to those who have kept the history alive. Denver has grown up considerably since its days as a railroad town, but it has a fantastic Colorado History Center, just down from the state capitol building, and a number of historic locomotives and passenger cars are on view at the Colorado Railroad Museum, twenty miles west in Golden. Colorado Springs, a hundred miles south, is much as it was. It is a divine experience to visit Glen Eyrie in the Rocky foothills outside of town, and not just because it is now a Christian conference center. Visitors can stay, as I did, in the castle that Palmer built for his Queen, and they can climb up into the ruddy-hued sandstone boulders of the Garden of the Gods, as I did, too. The town of Colorado Springs itself has the Pioneers Museum, which displays a bounty of General Palmer memorabilia, like the gold wedding ring he gave Queen. There is far more to see at sky-high

Leadville. The Tabor Opera House is open for tours, as are a dozen historic houses; and several museums, of which the most impressive is the National Mining Hall of Fame and Museum, with its walk-through re-creation of a silver mine among many other historic displays. The Tabors' Matchless Mine is just outside of town, still waiting to kick out silver. I took a tour there with the crackerjack guide Brenda Miller, who showed me the apparatus for hauling out carbonate and the crude shack where Baby Doe died, its interior lined with newspaper to keep out the cold.

My only actual train experience came at Cañon City, where I boarded the last vestige of the Rio Grande—a tourist train that runs up through the Royal Gorge, offering views of the forts high up on either side. Farther south, I was moved to see the mesas rising off to my left, and the last of the towering Rockies to my right, but, sadly, there is little in Pueblo and Cuchara to remind anyone of what they'd been, and even less in El Moro. Indeed, when I tried to find the original town, Google Maps directed me only to a nondescript intersection in a residential area that is now a quiet suburb of Trinidad. Sic transit gloria El Moro!

The Santa Fe's Raton Pass remains intact, but it's not so easy to find, since the train route over it is well away from the I-25. I had to pull off the highway, and then venture through some bushes and tall grass—past a sign indicating that this was still Wootton property, remarkably, all these years later—and on perhaps another quarter-mile, my fears of rattlesnakes rising, until I spotted the train tracks winding up from the valley below to reach a slight indent on the ridge above. From there, the tunnel was not quite visible.

Safely back in my rental car, I continued on to thriving Santa Fe, its adobe character derived from Fred Harvey's Fonda Hotel, and very much a tourist draw just as Harvey had hoped. Its historical museum pays homage to the Harvey Girls among many other celebrities and incidents of a former day, and adjoins the very Palace of the Governors where Don Miguel Otero secured a railroad charter for Strong's Santa Fe.

Kansas is not so bountiful for a train enthusiast, but Dodge City is a trip, what with its Boot Hill cemetery and an aging Santa Fe locomotive still on the tracks that split the town in two. I had a sublime moment driving back east from there to a motel in Newton, likewise a creation of the Santa Fe, on my way to Topeka, where the Santa Fe began. I'd gotten started late, and, nearly alone on the road as the night sky over the empty prairie turned jet black, I played some Chopin nocturnes loud on the car stereo, and, although by rights I should have been playing country music, they made a nearly holy soundtrack in the silence all around me.

On my travels I spent a great deal of time in the archives of the two rail-road lines, and it is to their selfless curators that I owe my greatest debt of gratitude. A treasury of Rio Grande documents is housed at History Colorado in Denver, where a pair of diligent librarians, Sarah Gilmore and Kerry Baldwin, patiently assisted me in my research. The Santa Fe equivalent is at the Kansas Historical Center in the grand Kansas Museum of History in Topeka. And there, the veteran archivist Lynn Frederickson expertly guided me through the railroad's immense holdings, of which the highpoint was a fat scroll she found for me. Unspooled, it rolled out a twenty-foot-long map of the Santa Fe's route from Topeka to Guaymas, depicting the elevations of the many stations along the way, their exact heights carefully inscribed in a perfect hand.

All historians depend on the historians who have gone before, and I'd like to single out for special praise the three historians who set themselves to the exhausting task of combing through mountains of corporate records to compile detailed business histories of my two railroads. I found their work indispensable. Such tireless researchers rarely get their due, and I would like to pay tribute to them now: Glenn D. Bradley, author of *The Story of the Santa Fe*, published a century ago in 1920; Keith L. Bryant, Jr., who wrote his *History of the Topeka & Santa Fe Railway* in 1974; and Robert G. Athearne, whose very readable history of the Rio Grande, *Rebel of the Rockies*, came out in 1962. Of them, sadly, only Bryant still lives. I salute you, sir! I'm also grateful to John S. Fisher for his evocative biography of General Palmer, *A Builder of the West*, from 1942. It's tragic that a companion biography of Strong has not been written, and probably never will be.

Solitary as it is, writing is never done entirely alone. On the business side, I have had a true and loyal friend in my longtime agent Dan Conaway of Writers House. For the editing side, I need to thank my valiant editor, Jofie Ferrari-Adler, the captain now of his own esteemed imprint, Avid Reader Press, who saw the literary possibilities in a forgotten railroad war, and has worked painstakingly to craft them into the story that is this book. To ease the text into print, his colleague Julianna Haubner stepped in to do a final, brilliant edit and oversee the copious production details. Anthony Newfield did the meticulous copyediting, and Kathryn Higuchi and the rest of the production team shepherded the book through the intricacies of that world. The sparkling Jordan Rodman and Morgan Hoit handled its national promotion. The ingenious Pete Garceau designed the darling cover; the discerning Jeffrey L. Ward created the many maps; and the ever-diligent Carol Poticny collected all the images that adorn this book. Many, many thanks to all.

And kudos to the remarkable Hannah Assadi, a talented novelist who took on the onerous task of organizing the book's endnotes and caught and killed more than a few errors in the process. (Any that remain are all on me.) A diligent Harvard undergrad, Eli Frankel, chipped in with some essential research, as did the learned train buff Bob Walz. For editorial advice and moral support I turned to my literary posse—Don Cummings, David Thomas, Lisa Dierbeck, Tony Kahn, Laura Strausfeld, David Frankel, Don Armstrong, and Pete Karmel. But I relied most heavily on Patrick McGrath, eminent novelist and dear pal who listened with patient, loving interest as I related this western saga over Rioja at more than one of Manhattan's better taverns. Irish by birth, Patrick is the most patriotic American I know, which is why I dedicate this story of America to him.

I can't close without a *salut d'amour* to my two extraordinary daughters, Sara and Josie, with boundless gratitude for the many blessings they have bestowed on me, of which possibly the most blessed are their delightful children—my grandchildren, as I think of them—Logan, Kyla, and Neva. I'm very keen on my dazzling stepchildren, Darya and Alex, too. But my greatest thanks, as always, goes to their mother, the writer Rana Foroohar, my beloved seatmate for the grand train ride of marriage, her on the aisle and me by the window.

Notes

INTRODUCTION: A VERY PERSONAL WAR

5 *The first was at the General's castle:* Athearn, *Rebel of the Rockies*, 51–52.

5 *To him, the jagged, snow-capped Rockies:* Ibid., 8.

6 *Ultimately, while he conceded:* Fisher, *A Builder of the West*, 74.

6 *After Antietam, he volunteered:* Ibid., 75–125.

8 *It was fitting that he:* Burke, *On the Sublime*, ed. J. T. Bolton, 58.

9 *He made Colorado Springs:* Sprague, *Newport of the Rockies*, 23–26.

9 *At that point, the Union Pacific:* Grodinsky, *Transcontinental Railway Strategy*, 1.

10 *The Santa Fe Railroad was owned:* Bradley, *The Story of the Santa Fe*, 95.

10 *That was never Palmer's way:* Fisher, *A Builder of the West*, 17.

11 *Reflecting the intimacy:* Ferrell, *Denver & Rio Grande*, 20.

11 *Barreling out of Kansas:* Bradley, *The Story of the Santa Fe*, 94–95.

11 *The General, on the other hand:* Fisher, *A Builder of the West*, 243.

14 *In March of 1876:* Bradley, *The Story of the Santa Fe*, 91.

15 *All, that is, except for the bracing determination:* Ibid., 95.

17 *He started in with the railroads:* Ibid., 95.

19 *Broadly, Strong spoke in actions:* Ibid., 166.

19 *When Palmer later reported:* Athearn, *Rebel of the Rockies*, 52.

20 *In 1877, the General had planted:* Ibid., 43.

21 *There was much in play:* Athearn, *Rebel of the Rockies*, 74.

PART ONE: THE RATON PASS

CHAPTER 1: ON THE TRAIN TO EL MORO

25 *Shortly before seven o'clock:* Borneman, *Rival Rails*, 132.

25 *A modest settlement:* Athearn, *Rebel of the Rockies*, 21.

25 *The Rio Grande train was most likely pulled:* Ferrell, *Denver & Rio Grande*, 30.

25 *It had been built back East:* Llanso, "Denver & Rio Grande", https://www
.steamlocomotive.com/locobase.php?country=USA&wheel=2-6-0&railroad
=drgw#11704.

26 *The fully reconstructed Las Animas:* Ferrell, *Denver & Rio Grande*, 30–31.

27 *The first steam-powered engines:* Wolmar, *The Great Railroad Revolution*, 19–22.

27 *The Rio Grande train lumbering:* Noble, *From the Clyde to California*, 124.

27 *The narrow cars could still:* Lucius Beebe & Charles Clegg, Rio Grande: Mainline of the Rockies, 11, complete with a dramatic photograph of the capsized train.

27 *At either end:* Stevenson, *From the Clyde to California*, 124.

27 *They were the chief engineers:* Wilson, *The Denver and Rio Grande Project*, 34.

27 *By early 1878, he'd become:* Ibid., 30–31.

27 *Shortly after leaving Glen Eyrie:* Borneman, *Rival Rails*, 130.

CHAPTER 2: THE WILD WEST

28 *The left-hand page:* Stegner, *Beyond the Hundredth Meridian*, 2.

34 *The United States, of course, was originally made:* "The Mexican American War," https://history.state.gov/milestones/1830-1860/texas-annexation.

34 *We have an unknown distance:* Brands, H. W. American Colossus, 557.

35 *May 10, 1969, was a glorious day:* Dodge, How We Built the Union Pacific Railroad, 29.

35 *A telegraph operator had:* Ambrose, *Nothing Like It in the World*, 366.

36 *A railroad to the Pacific:* Ibid., 108.

36 *Lincoln had also set:* Hilton, "A history of track gauge," https://trn.trains.com/railroads/abcs-of-railroading/2006/05/a-history-of-track-gauge.

37 *Winter snows made the line:* White, *Railroaded*, 85. In offering this citation, I need to add that I owe much of my skepticism of the first transcontinental to Richard White, whose delightfully contrarian *Railroaded* goes into compelling detail about the exaggerations of the Pacific Railway's achievement. I note that even the two magnificent accounts of the building of the railroad, Stephen Ambrose's *Nothing Like It in the World*, and David Haward Bain's *Empire Express* (both of them, curiously, published in the same year, 1999), make precious few claims for its utility or commercial value after devoting hundreds of pages to its construction.

38 *By 1869, there were plenty:* "Chronology of Railroading in America," https://www.aar.org/data/chronology-railroading-america/.

38 *Why Council Bluffs?:* Dodge, How We Built the Union Pacific Railroad, 57.

39 *Sympathetic river men:* Jackson, *Rails Across the Mississippi*, 4.

39 *The lawyer for the Chicago and Rock:* Pfeiffer, "Bridging the Mississippi," accessed online at https://www.archives.gov/publications/prologue/2004/summer/bridge.html.

40 *After the first transcontinental:* Ambrose, *Nothing Like It in the World*, 369.

41 *And they did, ultimately running:* "Railroads in the late 19th. Century," http://www.loc.gov/teachers/classroommaterials/presentationsandactivities/presentations/timeline/riseind/railroad/.

41 *At the high end, the wealthy:* Richter, "At Home Aboard" in Gender and Landscape, 77.

41 *The first car held:* Joyce B. Lohse, *General William Palmer: Railroad Pioneer*, 66.

42 *So observed Ralph Waldo Emerson:* Emerson, *The Journals and Miscellaneous Notebooks of Ralph Waldo Emerson*, 296.

43 *Far ones remained fixed:* Schivelbusch, *The Railway Journey*, 63.

43 *Nine years after:* Stilgoe, *Metropolitan Corridor*, 250.

43 *"I get so bored . . .":* Schivelbusch, *The Railway Journey*, 58.

44 *Freud saw the erotic:* Ibid., 77.

44 *It was better:* Richter, *Home on the Rails*, 59–61.

45 *One eastern journalist:* Schwantes, *The Pacific Northwest*, 193.

45 *A distraught railroad agent:* Alden, "Air Towns and their Inhabitants," 829.

46 *While that promotional effort:* Quiett, *The Built the West*, 84–85.

46 *Apparently, he'd felt insulted:* Train, *My Life in Many States and in Foreign Lands*, 293.

46 *He drinks no spirits:* Albert Dean Richardson, *Beyond the Mississippi*, 565.

47 *After crossing the broad:* Bain, *Empire Express*, 292.

CHAPTER 3: BEWARE THE PRAIRIE LILIES

48 *He was with the Kansas Pacific:* Fischer, *Builder of the West*, 149–150.

48 *The idea was to take:* Ibid., 176.

48 *Among the applicants:* Sprague, *Newport in the Rockies*, 17–18.

49 *As he journeyed, Bell was stunned:* Pierce, *Making the White Man's West*, 166.

50 *It took a few months:* Higgins, *To California over the old Santa Fe Trail*, 18.

50 *He was intrigued by some:* Bell, *New Tracks in North America*, 98–106.

51 *Sleeping so close by:* Fisher, *Builder of the West*, 136.

51 *The visit made for such a riot:* Ibid., 136.

51 *To the General, Bell was:* Sprague, *Newport in the Rockies*, 17.

52 *He learned that each cut:* Ibid., 64.

53 *On either side, Palmer later wrote:* Bell, *New Tracks in North America*, 175.

53 *Up on the surrounding desert:* Ibid., 176.

54 *How we got up, God knows:* Ibid., 178–179.

55 *Lucien Bonaparte Maxwell was a lively:* Inman, *Old Santa Fé Trail*, chapter XVIII, unpaginated.

55 *The history of the Maxwell grant:* Montoya, *Translating Property*, 52–53.

55 *For years, he lived on the estate:* Ibid., 70.

55 *Overwhelmed by it all, Maxwell:* Inman, *The Old Santa Fé Trail*, 300.

56 *After Maxwell got in touch:* Murphy, *Lucien Bonaparte Maxwell*, 184.

CHAPER 4: THE GRID

57 *Early on, the deal was even:* "Railroads, Federal Land Grants to," https://www
.encyclopedia.com/history/encyclopedias-almanacs-transcripts-and-maps
/railroads-federal-land-grants-issue.

57 *Of the federal grants:* "Atchison, Topeka, & Santa Fe Railroad Co. Land Grant
Records," https://www.kshs.org/p/atchison-topeka-santa-fe-railroad-co-land-records
/19951.

58 *The federal land was doled:* "Railroads, Federal Land Grants to," https://www
.encyclopedia.com/history/encyclopedias-almanacs-transcripts-and-maps
/railroads-federal-land-grants-issue.

58 *Everywhere I hear the sound:* Simonin, *The Rocky Mountain West in 1867*, 57.

59 *Like so much else in America:* Stanislawski, "The Origin and Spread of the Grid-Pattern Town," 105–120.

59 *It was brought by the early:* Hudson, "Towns of the Western Railroads," 41–52.

60 *In 1873, a director:* Reps, *Cities of the American West: A History of Frontier Urban Planning*, 568.

61 *The tracks usually ran:* Hudson, "Towns of the Western Railroads," 47.

61 *All newcomers arrived:* Schwantes, *The West the Railroads Made*, 153.

61 *While the train tracks:* Hudson, "Towns of the Western Railroads," 46.

CHAPTER 5: WHERE TO GO

63 *The standard text:* Wellington, *The Economic Theory of the Location of the Railroads*, 1.

64 *To Wellington's way of thinking:* Ibid., 731.

64 *Born in New York:* Sprague, *Newport in the Rockies*, 27.

65 *Coming south from Denver:* Sprague, *Newport in the Rockies*, 28.

65 *In Cañon City, voters:* Ibid. 26.

65 *Its citizens doubled:* Ibid., 23.

65 *The contract required:* Ibid., 25.

65 *It was irksome, but South Pueblo:* Whitaker, *Pathbreakers and Pioneers of the Pueblo Region*, 113.

66 *This time, the General turned:* Athearn, *Rebel of the Rockies*, 25–26.

66 *By 1873, the railroads had succeeded:* Stiles, *The First Tycoon*, 568.

67 *By 1873, the total railroad investment:* Bellesiles, *1877*, 2.

67   *"If I had been struck . . .":* Ibid., 3–4.

67 *Five thousand businesses:* Ibid., 4–6.

68 *This one, the Panic of 1873:* "The Panic of 1873", https://www.pbs.org/wgbh/americanexperience/features/grant-panic/.

68 *This was the downside:* Bellesiles, *1877*, 114.

69 *A certain "FBH" wrote:* Athearn, *Rebel of the Rockies*, 34–35.

69 *The Gazette's editors rushed:* Ibid., 30.

71 *He recognized that railroads:* Ibid., 48.

71 *now a thriving:* Colorado Chieftain, August 16, 1876, accessed at https://www.coloradohistoricnewspapers.org/cgi-bin/colorado?a=d&d=CFT18771108-01.2.18&e=mg-txIN%7ctxCO%7ctxTA.

72 *Five miles short:* Athearn, *Rebel of the Rockies*, 39–43.

72 *The* Colorado Chieftain: Lavender, *The Southwest*, 262.

73 *Still, to be on the safe side:* Fisher, *A Builder of the West*, 247.

CHAPTER 6: SEEKING UNCLE DICK

74 *While the Maxwell Grant began:* Borneman, *Rival Rails*, 132.

74 *The son of a Kentucky planter:* Conrad, *"Uncle Dick" Wootton*, 30–50.

74 *In 1852, he realized:* Ibid., 250–277.

75 *In 1867, he saved Maxwell's life:* Murphy, *Lucien Bonaparte Maxwell*, 84.

75 *Two of the more memorable visitors:* Conrad, *"Uncle Dick" Wootton*, 436–438.

77 *He pushed so hard:* Bradley, *The Story of the Santa Fe*, 98.

77 *He directed his chief engineer:* Osterwald, *Rails thru the Gorge*, 43.

77 *He'd worked himself:* Ibid., 42.

77 *He'd directed a horde:* Bradley, *The Story of the Santa Fe*, 51.

77 *James McMurtrie of the Rio Grande:* Osterwald, *Rails thru the Gorge*, 35.

77 *His technical skill was superb:* "A Mountain Holiday," *Lippincott's Magazine*, 534.

CHAPTER 7: SANTA FE, INC.

79 *It was the brainchild:* Bradley, *The Story of the Santa Fe*, 25–26.

80 *He arrived as an abolitionist:* "Letters of Cyrus Kurtz Holliday," https://www.kshs .org/p/letters-of-cyrus-kurtz-holliday-1854-1859/12717.

81 *"God might have made . . .":* Ibid.

81 *They settled on a strip:* Bradley, *The Story of the Santa Fe*, 27.

81 *"When I commenced . . .":* "Letters of Cyrus Kurtz Holliday," https://www.kshs.org /p/letters-of-cyrus-kurtz-holliday-1854-1859/12717.

81 *Malaria and cholera:* Goodrich, *War to the Knife*, 66–69.

81 *He also helped organize:* Bradley, *The Story of the Santa Fe*, 27.

82 *Undeterred, Holliday organized:* "Letters of Cyrus Kurtz Holliday," https://www .kshs.org/p/letters-of-cyrus-kurtz-holliday-1854-1859/12717.

82 *In 1859, he became a delegate:* Treadway, *Cyrus K. Holliday*, 138–139.

82 *After he won a state charter:* Bradley, *The Story of the Santa Fe*, 34–35.

83 *"The coming tides . . .":* Bradley, *The Story of the Santa Fe*, 45–46.

83 *The local newspaper thought:* Ibid., 46.

83 *He added Santa Fe:* Ibid., 47–51.

83 *It needed moneymen:* Bradley, *The Story of the Santa Fe*, 47–48.

83 *From this esteemed collective:* Coolidge, *The Autobiography of T. Jefferson Coolidge*, 1–2.

84 *When he returned:* Ibid., 88.

85 *"I resigned as soon . . .":* Ibid., 88.

85 *He failed to mention:* Bryant, *Atchison, Topeka and Santa Fe Railway*, 75.

85 *An early point of tension:* Bradley, *The Story of the Santa Fe*, 79.

86 *"I thought of the poor families . . .":* Ibid., 77–78.

86 *Sure enough, he put up:* Ibid., 79.

86 *If any town in America:* Bryant, *Atchison, Topeka and Santa Fe Railway*, 32–33.

87 *In Dodge, the wrong side:* Clavin, *Dodge City*, 25–28.

87 *To beef it up, the railroad authorized:* "Wyatt Earp dropped from Wichita Police Force," https://www.history.com/this-day-in-history/wyatt-earp-dropped-from -wichita-police-force.

88 *The Santa Fe decided to revise:* DeArment, *Bat Masterson*, 71.

89 *When a crew of rowdy Texas:* Bradley, *The Story of the Santa Fe,* 64.

89 *Typical of the Boston Crowd:* Ibid., 44.

CHAPTER 8: ENTER THE QUEEN

90 *In January 1869, the General was aboard:* Fisher, *A Builder of the West,* 151.

90 *After parting with her at Cincinnati:* Sprague, *Newport in the Rockies,* 20–21.

91 *Although Mellen had not been implicated:* Gehling, *Queen & Her General,* loc. 37, Kindle.

92 *She grew into a fiercely independent:* Black, *Queen,* 13–15.

92 *"To Miss Queen Mellen . . .":* Ibid., 19.

93 *It started when he thanked:* Ibid., 21.

93 *"I do not believe much in confessions . . .":* Ibid., 22.

94 *While Palmer inquired: Letters 1853–1868 Gen'l Wm. J. Palmer,* compiled by Isaac H. Clothier, 22.

94 *The word "homosexual" did not exist:* Lowry, *Story the Soldiers Wouldn't Tell,* loc. 1642, Kindle.

94 *Clothier, curiously, had made a similar complaint:* Gehling, *Queen & Her General,* loc. 80–84, Kindle.

95 *The spring after he proposed:* Ibid., loc. 105–108, Kindle.

95 *Palmer did his best to assure:* Black, *Queen,* 24.

95 *"Is it any wonder . . .":* Ibid., 26.

96 *"I could not sleep any more . . .":* Fisher, *Builder of the West,* 162.

96 *He bathed in the chilly waters:* Ibid., 162.

96 *Before the week was up:* Ibid., 190.

96 *Before the week was up:* Sprague, *Newport in the Rockies,* 33.

96 *He planned to line them:* Ibid., 30.

97 *It would be a club:* Ibid., 30.

97 *There together amid:* Ibid., 28–29.

98 *That fall, the General invited:* Fisher, *A Builder of the West,* 165–166.

98 *A "little railroad," as he put it:* Ibid., 177–178.

98 *"I felt so happy that . . .":* Black, *Queen,* 31.

99 *She must not have shared:* Ibid., 32.

99 *It wasn't until April:* Ibid., 34.

100 *The worst realization:* Black, *Queen,* 34.

100 *Her uncle Malcolm Clarke:* "Montana History Almanac," https://missoulian.com/lifestyles/territory/montana-history-almanac-trader-killed-for-alleged-insult-to-piegan/article_48c9b0a0-c51d-11e0-ae35-001cc4c002e0.html.

100 *Together, she and Palmer:* Ibid., 34–35.

101 *He came brightly gartered:* Ibid., 35.

101 *When Queen returned to Flushing:* Fisher, *A Builder of the West,* 185.

101 *Her one surviving letter:* Black, *Queen,* 243.

102 *The service was performed:* Ibid., 37.

102 *She'd gone to bed:* Ibid., 40–43.

103 *"You may imagine Colorado Springs . . .":* Kingsley, *South by West,* 47–48.

103 *That winter, temperatures rarely rose:* Ibid., 48–53.

104 *To give her some space:* Sprague, *Newport in the Rockies,* 43–44.

104 *"Our mood was a very happy . . .":* Blevins, ed., *Legends, Labors & Loves,* 274.

104 *Palmer let Elsie stay:* May, *A Kingdom of Their Own,* 66.

105 *A gossipy New York sophisticate:* Wolcott, *Heritage of Years,* 80.

105 *Wolcott spotted Queen:* Ibid., 99.

105 *"Queen Palmer climbing . . .":* Ibid., 88.

106 *Before long, Queen took over:* Ibid., 88.

106 *Wolcott's account suggested:* Ibid., 74.

106 *"I was in a furious . . .":* Sprague, *Newport in the Rockies,* 45.

106 *Unable to find the right:* Black, *Queen,* 43–44.

107 *Ultimately, she abandoned:* Ibid., 44.

107 *But, unable to face:* Ibid., 44.

107 *When the visit to Paris:* Ibid., 44.

107 *Palmer was all alone:* "Glen Eyrie," https://coloradoencyclopedia.org/article/glen
      -eyrie.

CHAPTER 9: THE BATTLE IS JOINED

108 *Just twenty-six, Morley was:* Cleaveland and Fitzpatrick, *The Morleys,* 41.

109 *After the war, Morley fell:* Ibid., 59–63.

109 *By then, Morley had married:* Ibid., 2–3.

110 *He'd go about his surveying:* Ibid., 162.

110 *When he revealed his plan:* Ibid., 160.

111 *He told Robinson and Morley:* Bryant, *Atchison, Topeka and Santa Fe Railway,* 46.

111 *There was nothing like it:* Ibid., 46.

112 *On December 17, 1877, Morley's work:* Bradley, *The Story of the Santa Fe,* 107.

112 *The news Strong received:* Ibid., 96.

113 *It seemed the SP's legislation:* Ibid., 97.

113 *This posed a challenge:* Ibid., 97.

113 *He was still inclined to go west:* Athearn, *Rebel of the Rockies,* 44.

114 *"They will try and get a force . . .":* "Raton Pass" Letters, February 25–28, 1878,
      Letters of Albert A. Robinson, The Archives of the Atchison, Topeka and Santa
      Fe Railroad.

114 *He fired off a coded message:* Bradley, *The Story of the Santa Fe,* 100.

115 *When the two men disembarked:* It needs to be said here that the events of
      that night are clouded in some mystery, as the various historical accounts—
      even the near-contemporary ones—rarely square. In the aggregate, they leave
      considerable uncertainty as to various details regarding timing, participants, and
      specific acts, although there is little disagreement about the basic contours of
      the incident, and none about the outcome. So it is left to me, ultimately, to

make my best guess as to what happened. In crafting this account, I have relied on the facts that everyone agrees on, and then applied common sense to decide the issues that are in dispute. By common sense, I mean my understanding of the essential motivations of the participants. Where the factual truth simply cannot be established—in this conflict and in others to come in this narrative— I have gone with the emotional truth of the participants' core motivations as I understand it. In the end, I'm guessing that people are most likely to act like themselves.

116 *Morley had already assembled:* Bradley, *The Story of the Santa Fe,* 100.

116 *There, Robinson handed Uncle Dick:* "Raton Pass" Letters, February 25–28, 1878, Letters of Albert A. Robinson, The Archives of the Atchison, Topeka and Santa Fe Railroad.

117 *And not just common laborers:* Ibid.

117 *The test came before dawn:* Bradley, *The Story of the Santa Fe,* 100.

117 *"They found they were . . .":* "Raton Pass" Letters, February 25–28, 1878, Letters of Albert A. Robinson, The Archives of the Atchison, Topeka and Santa Fe Railroad.

118 *The only route into Leadville:* Bradley, *The Story of the Santa Fe,* 100–103.

PART TWO: THE ROYAL GORGE

CHAPTER 10: PRECIOUS METALS

124 *At six cents a pound:* Blair, *Leadville: Colorado's Magic City,* 21.

124 *In 1874, when Wood and Stevens:* "The Price of Gold, 1257–Present," https:// www.measuringworth.com/datasets/gold/result.php?goldsilver=on&year_source =1687&year_result=2020.

124 *By the Coinage Act of 1873:* "US Mint History," https://www.usmint.gov/news /inside-the-mint/mint-history-crime-of-1873.

124 *The eastern financiers wanted:* Ibid.

125 *In 1878, a panicked Congress:* "Cross of Gold Speech", http://historymatters.gmu .edu/d/5354/.

125 *The issue so gripped:* "President McKinley Signs Gold Standard Act," https://www .politico.com/story/2013/03/this-day-in-politics-088821.

125 *Wood and Stevens kept quiet:* Blair, Leadville: *Colorado's Magic City,* 22–23.

126 *This time, the quality so impressed:* Ibid., 38.

126 *A silver mine is not just:* The National Mining Hall of Fame and Museum in Leadville, Colorado, has a wonderful scale model along with a walk-through exhibit and countless other artifacts of this hazardous line of work if anyone is curious to see what silver mining was like in the late nineteenth century.

126 *No wagons could carry:* Blair, Leadville: *Colorado's Magic City,* 23.

126 *Wood and Stevens set the men's wages:* Ibid., 23–24.

127 *Unnerved by the incident:* Ibid., 27.

127 *Described by the* Chronicle: *Leadville Daily, The Evening Chronicle*, https://www
    .coloradohistoricnewspapers.org/?a=d&d=LEC18790308-01.1.1&.

127 *Palmer had been following:* Athearn, *Rebel of the Rockies*, 53.

127 *In a follow-up report:* Griswold, *The Carbonate Camp Called Leadville*, 134–137.

127 *"With a railroad . . .":* Ibid., 136.

129 *For now, the General pictured:* Waters and Malott, *Steel Trails to Santa Fe*, 103.

129 *They quietly established:* Bryant, *Atchison, Topeka and Santa Fe Railway*, 47–50.

CHAPTER 11: "A GAME OF BLUFF"

130 *"They are determined . . .":* Griswold, *The Carbonate Camp Called Leadville*, 138.

130 *The Denver papers were:* Borneman, *Rival Rails*, 139.

131 *McMurtrie wasn't so sure:* Ibid., 134–139.

131 *Unwilling to commit:* Athearne, *Rebel of the Rockies*, 56.

132 *Once the Santa Fe crews:* Borneman, *Rival Rails*, 146.

132 *Sure enough, early on the morning:* Ibid., 146.

132 *"SEE TO IT WE DO NOT . . .":* Ibid., 146.

133 *Trapped at the station:* Bryant, *Atchison, Topeka and Santa Fe Railway*, 48.

134 *In Pueblo, he'd stabled:* Cleaveland and Fitzpatrick, *The Morleys*, 169.

134 *The* Chieftain *reported:* Ibid., 171.

134 *When his train reached Cañon City:* Ibid., 171.

134 *Friendly to the Santa Fe:* Ibid., 172.

CHAPTER 12: HAW

136 *There was another silver strike:* Osterwald, *Rails thru the Gorge*, 44.

136 *Once a rowdy mining camp:* Blair, *Leadville: Colorado's Magic City*, 40.

137 *To him, though, his greatest accomplishment:* Burke, *The Legend of Baby Doe*, 39

137 *Born in Holland, Vermont:* Ibid., 40.

138 *To try to prove himself:* Ibid., 41–44.

139 *Soon enough, the Little Pittsburg:* Blaire, *Leadville: Colorado's Magic City*, 47–50.

139 *The other two men sold:* Ibid., 50.

140 *What had once been a town:* Paul, *Mining Frontiers of the Far West 1848–1880*, 128.

140 *It became such a destination:* Ibid., 128.

140 *"BRAWL IN A STATE . . .":* Burke, *The Legend of Baby Doe*, 61.

140 *"IS THERE NO . . .":* Griswold, *The Carbonate Camp Called Leadville*.

141 *Finally, Tabor settled . . . :* Burke, *The Legend of Baby Doe*, 60.

141 *He appointed himself:* Ibid., 61–62.

141 *Knowing of Leadville's interest:* Ellmann, *Oscar Wilde*, 204.

142 *Before long, a waiter:* Burke, *The Legend of Baby Doe*, 68–71.

143 *The champagne flowed:* Ibid., 72.

CHAPTER 13: THE DEAD-LINE

144 *He knew all about that claim:* Athearne, *Rebel of the Rockies*, 59.

144 *Worse, Palmer's history:* Ibid., 60–61.

145 *By night, they faced:* Bryant, *Atchison, Topeka and Santa Fe Railway*, 49–50.

145 *Reversing that now:* Osterwald, *Rails thru the Gorge*, 36.

146 *Up went DeRemer's men:* Cleaveland and Fitzpatrick, *The Morleys*, 174.

146 *Ray Morley was in charge:* Athearne, *Rebel of the Rockies*, 65.

147 *In his judgment:* Ibid., 63.

147 *Where there wasn't room:* Osterwald, *Rails thru the Gorge*, 45.

148 *"If this plan proceeds . . .":* Athearne, *Rebel of the Rockies*, 64.

148 *It bore the words:* Ibid., 63.

148 *Bluff or not, such boldness:* Ibid., 64.

149 *By now, the Santa Fe:* Ibid., 65.

149 *He offered to take:* Ibid., 66.

151 *But the day passed:* Ibid., 67–68.

152 *"[If] they were to put . . .":* Ibid., 68.

152 *"I did not think . . .":* Ibid., 67.

152 *On Friday, December 13:* Ibid., 68–69.

CHAPTER 14: THIS MEANS WAR

154 *He directed Robert Weitbrec:* Ibid., 72.

154 *He drove up the rates:* Bradley, *The Story of the Santa Fe*, 117.

155 *Initially, he had been confident:* Athearne, *Rebel of the Rockies*, 73.

155 *Unable to halt it:* Ibid., 72–74.

156 *To the* Rocky Mountain News: Bradley, *The Story of the Santa Fe*, 123.

156 *A resident of Pueblo:* Ibid., 124.

156 *"The complaints of injustice . . .":* Ibid., 125.

156 *In April, Palmer went:* Athearne, *Rebel of the Rockies*, 75.

157 *He hired gun-toting soldiers:* Bryant, *Atchison, Topeka and Santa Fe Railway*, 51.

158 *The days ticked by:* Athearne, *Rebel of the Rockies*, 75.

158 *In Palmer's Colorado Springs:* Ibid., 75.

158 *Nonetheless, the ruling:* Fisher, *A Builder of the West*, 258.

159 *The* New-York Tribune *observed:* Athearne, *Rebel of the Rockies*, 76.

159 *After splashy eastern publications:* Blair, *Leadville: Colorado's Magic City*, 55.

159 *As more miners came in:* "Leadville, Colorado," https://westernmininghistory.com /towns/colorado/leadville/.

159 *No court had blocked:* Bradley, *The Story of the Santa Fe*, 130.

161 *In 1864, an innovative:* "A Tribute to Sheriff Cook," https://www.arapahoegov .com/DocumentCenter/View/8440/A-tribute-to-Sheriff-Cook.

161 *When the Rio Grande:* "New Mexico's Lincoln County War," https://www .legendsofamerica.com/nm-lincolncountywar/.

CHAPTER 15: BAD MEN

163 *Gun tallies came in:* Athearne, *Rebel of the Rockies*, 77.

164 *"A splendid body . . .":* Ibid., 77.

164 *Details are spotty:* Ibid.,79.

164 *All of them had come:* Bradley, *The Story of the Santa Fe*, 124.

165 *An obscure Boston financier:* Athearne, *Rebel of the Rockies*, 74.

165 *A further wrinkle:* "Back Bay Houses," https://backbayhouses.org/79-marlborough/.

166 *To Strong's horror:* Bradley, *The Story of the Santa Fe*, 127.

167 *In a panic, Strong rushed:* Athearne, *Rebel of the Rockies*, 80.

167 *To enlist popular support:* Ibid., 79.

167 *To secure Pueblo:* Bradley, *The Story of the Santa Fe*, 127.

167 *Then he returned to the depot:* Athearne, *Rebel of the Rockies*, 80.

168 *In Denver, the sheriff:* Bradley, *The Story of the Santa Fe*, 128.

168 *What the papers called:* Ibid., 128.

168 *The many florid newspaper accounts:* Athearne, *Rebel of the Rockies*, 81. He is dismissive of Bat's involvement, per his long note, but most other accounts include him.

169 *Finally, Palmer recognized:* Fisher, *A Builder of the West*, 257.

169 *When another judge in the building:* Athearne, *Rebel of the Rockies*, 82–83.

169 *In the end, Judge Hallett did offer:* Ibid., 83.

CHAPTER 16: GOULD

172 *Just five feet tall, withered:* Klein, *The Life and Legend of Jay Gould*, 3.

172 *Born on a farm:* Renehan, *Dark Genius of Wall Street*, 72.

173 *He left for New York City:* Ibid., 83.

174 *There had been a stock exchange:* Pisani, "This Single Paged Document Started the New York Stock Exchange 225 Years ago," https://www.cnbc.com/2017/05/17/this-single-paged-document-started-the-new-york-stock-exchange-225-years-ago.html.

174 *"These men are the Nimrods . . .":* Brands, *American Colossus*, loc. 17, Kindle.

174 *As a sideline to the railroad:* Klein, *The Life and Legend of Jay Gould*,77–78.

175 *In his epic rise:* Stiles, *The First Tycoon*, 154.

175 *It was widely said:* Brands, *American Colossus*, 18.

175 *In desperation, Vanderbilt persuaded:* Renehan, *Dark Genius of Wall Street: The Misunderstood Life of Jay Gould, King of the Robber Barons*, 117–119.

175 *Drew fled to New Jersey:* Ibid., 124.

176 *When Gould sued to recover:* Ibid., 127–134.

176 *This is when he became:* Ibid., 179.

176 *He huddled there, nothing:* Ibid., 177.

176 *"It won't do, Josie . . .":* Ibid., 192.

176 *When Vanderbilt realized:* Ibid., 185.

176 *He used his winnings:* Ibid., 229.

176 *When he purchased the Pacific:* Morris, *The Tycoons*, 140.
178 *He declared that Boston:* "CB&Q Builds to Denver," https://history.nebraska.gov
     /sites/history.nebraska.gov/files/doc/publications/NH1959CBQtoDenver.pdf.
178 *In September 1879, after the:* Athearne, *Rebel of the Rockies*, 85.
179 *It would be the death:* Ibid., 86.
179 *Matters were made even worse:* Ibid., 87.
180 *To avoid future conflict:* Ibid., 87.

CHAPTER 17: A WHISKEY SALUTE

181 *By the terms of Gould's "Treaty . . .":* Ibid., 88.
181 *On July 22, 1880, the first:* Blair, *Leadville: Colorado's Magic City*, 151.
181 *One early passenger:* Ingersoll, *Crest of the Continent*, 194.
182 *Since the General was away:* Black, *Queen of Glen Eyrie*, 84.
183 *In her weakened state, Queen stayed:* Ibid., 84.
184 *To reach Santa Fe itself:* Bradley, *The Story of the Santa Fe*, 137–138.
184 *The governor of New Mexico:* Bryant, *Atchison, Topeka and Santa Fe Railway*, 62.
184 *Under Strong, the railroad:* Ibid., 63.

CHAPTER 18: HARVEY HOUSES

186 *The son of a destitute tailor:* Fried, *Appetite for America*, loc. 285–460, Kindle.
187 *"They'll try anything . . .":* Ibid., loc. 1051.
187 *A Topeka newspaper extolled:* Ibid., loc. 1093.
188 *Phillips had been top chef:* Ibid., loc. 1221–1228.
188 *A correspondent for a London sporting newspaper:* Ibid., loc. 1242.
188 *Phillips's restaurant was such a hit:* Ibid., loc. 1260–1317.
189 *The Montezuma's menu:* Ibid., loc. 160–166.
189 *Although located in the desert:* Ibid., 166–173.
189 *As the saying went:* Ibid., loc. 1720.
190 *Some knives came out:* Ibid., loc. 1692–1699.
190 *At the very least:* Ibid., loc. 1713.
191 *With his restaurants:* Bryant, *Atchison, Topeka and Santa Fe Railway*, 114–117.
192 *At some of his hotels:* Fried, *Appetite for America*, loc. 3282, Kindle.

PART THREE: LOS ANGELES

CHAPTER 19: THE PUEBLO

197 *The Central Pacific offered:* Nordhoff, *California: for Health, Pleasure, and Residence*, 28.
197 *By then, California:* James Bryce, *"The American Commonwealth"*
198 *As it grew, the CP became:* "Railroads Beginning in California," https://www
     .railswest.com/history/californiabeginnings.html.
198 *By 1880, the Southern Pacific:* Orsi, *Sunset Limited*, 20.
199 *To Oscar Wilde, it was:* Starr, *Americans and California Dream*, loc. 246, Kindle.

199 *He made straight:* Ibid., 246.

199 *By 1900 it had pulled:* Rodman Wilson Paul, *Mining Frontiers of the Far West, 1848–1880,* 27. He cites a figure of $1.3 billion as of 1900, which accounts for the sum in the text.

199 *"What crowds are rushing . . .":* Colton, *The Land of God,* 358.

200 *One might be able:* Famously, in the *Wealth of Nations.*

200 *By the time Strong and the General started:* Starr, *Americans and the California Dream,* 391.

200 *The Southern Pacific line:* Quiett, *They Built the West,* 271.

201 *A typical homesteader could sell:* McWilliams, *Southern California: An Island on the Land,* 115d.

201 *One of the more notable:* Starr, *Americans and the California Dream,* 200.

202 *It was founded in 1874 by a small group:* Scheid, *Pasadena: Crown of the Valley,* 31–32.

202 *Initially, the sick visitors:* Ibid., 44.

202 *In 1880, the downtown:* Ibid., 46.

CHAPTER 20: THE BIG FOUR

203 *"I never had any idea . . .":* Rayner, *The Associates: Four Capitalists Who Created California,* loc. 124, Kindle.

203 *Even in its earliest incarnation:* McWilliams, *California,* 179.

203 *It was said that before:* Lavendar, *The Great Persuader,* 330.

204 *"His feet are more often . . .":* Rayner, *The Associates,* loc. 1351, Kindle.

204 *He was best known for:* Ibid., loc. 1296–1303.

204 *He had a touch of public-spiritedness:* Ibid., loc. 1907.

204 *Born to a broken-down farmer:* Ibid., loc. 41–138.

205 *The first to try was Tom Scott:* Ibid., loc. 880–890.

206 *Huntington hired a political fixer:* Ibid., loc. 1402.

206 *"He was a little cross . . .":* Ibid., loc. 1475–1483.

207 *Poole brought an SP man:* Rayner, *The Associates,* loc. 1592–1647, Kindle.

207 *Fifty ranchers thundered up:* Ibid., loc 1670–1690.

208 *It was why Frank Norris:* Norris, *The Octopus,* 63.

CHAPTER 21: "A TERRIBLE, SINGLE-HANDED TALKER"

209 *By the time of the San Joaquin Valley:* "Mussel Slough Incident," https://www.encyclopedia.com/history/dictionaries-thesauruses-pictures-and-press-releases/mussel-slough-incident.

209 *In 1880, it looked like:* Bryant, *Atchison, Topeka and Santa Fe Railway,* 95.

209 *For the purpose, they turned:* Abramo, "A guide to the Kimball Family Collection," http://www.nationalcityca.gov/home/showdocument?id=145.

210 *He replied with his famous line:* Smythe, *History of San Diego,* 284.

210 *Kimball could be:* Ibid., 285.

210 *Still, he had offered the Santa Fe:* Bryant, *Atchison, Topeka and Santa Fe Railway*, 97.

210 *The first shipment was delivered:* Ibid., 97.

212 *When the first train rolled in:* Ibid, 99.

212 *By 1884, the California Southern:* Ibid., 100.

CHAPTER 22: GUAYMAS

213 *By the end of 1880, he'd gotten:* Bradley, *The Story of the Santa Fe*, 139.

216 *Before the week was out:* Ibid., 178–179.

216 *If he couldn't build out:* Bradley, *The Story of the Santa Fe*, 152–153.

217 *Back in 1878, he had sent:* Bryant, *Atchison, Topeka and Santa Fe Railway*, 80.

218 *Just as the California Southern had:* Boyd, *20 years to Nogales*, 297.

218 *After a San Francisco–based paper:* Ibid., 308.

219 *For his part, Strong declared:* Ibid., 308.

219 *While A. A. Robinson was officially:* Cleaveland, *The Morleys*, 199–200.

219 *To cheer him up, Ada sent:* Ibid., 202-203.

219 *Stretching back over seventeen hundred miles:* Bradley, *The Story of the Santa Fe*, 153.

219 *The combination also created:* Bryant, *Atchison, Topeka and Santa Fe Railway*, 82.

CHAPTER 23: "THIS IS HARD"

221 *While that had long been his quest:* Fisher, *A Builder of the West*, 278.

222 *The General had first visited Mexico:* Pletcher, "General William S. Rosecrans and the Mexican Transcontinental Railroad Project," 663.

222 *As a precaution, the General:* Black, *Queen of Glen Eyrie*, 73–74.

222 *The Palmer party landed:* Fisher, *A Builder of the West*, 217–218.

222 *Palmer had by then acquired:* Ibid., 220.

223 *He raced to New York to meet:* Brandt, *The Railway Invasion of Mexico*, 79.

223 *Strong forcefully rebutted:* Ibid., 79–80.

223 *He noted that Palmer was known:* *Railway Review*, Aug 7, 1880.

224 *On September 8, 1880, Díaz gave:* Pletcher, "Mexico Opens the Door to American Capital, 1877–1880", 12.

225 *"A man can build . . .":* Warman, *The Story of The Railroad (Illustrated)*, loc. 3481, Kindle.

225 *At the height of the building:* Ibid., loc. 3496.

225 *He soon boosted:* "Historic Texas County Population," https://txcip.org/tac/census /hist.php?FIPS=48141.

225 *He sat there stricken:* Cleaveland with Fitzpatrick, *The Morleys*, 212.

226 *For the rest of his days:* Ibid., 217.

226 *He completed the line from El Paso:* Warman, *The Story of the Railroad*, loc. 3459, Kindle.

CHAPTER 24: THE A&P

227 *The burning desert plateau:* Railway Review, May 13, 1892, 276.

228 *Palmer banked with the Monte de Piedad:* Bancroft, History of Mexico 1883–88, 557.

228 *Although the company stock:* Bryant, Atchison, Topeka and Santa Fe Railway, 82.

229 *It was only a third:* Ibid., 84.

229 *In calculating the costs:* Goetzmann, Exploration and Empire, 289.

230 *He organized a railroad company:* Bryant, Atchison, Topeka and Santa Fe Railway, 84.

230 *Soon after, a brand-new:* Ibid., 85.

230 *While Whipple had plotted:* Ibid., 90.

231 *Water was so scarce:* Ibid., 86.

232 *Soon, Kingman hit another obstacle:* Ibid., 89–90.

232 *Six workers were crushed:* Borneman, Rival Rails, 211.

232 *To defeat him, Huntington would build:* Bryant, Atchison, Topeka and Santa Fe Railway, 91.

232 *The deal was struck:* Ibid., 92.

233 *The appointment made Strong:* Ibid., 76.

234 *While the Big Four were competing:* Files of the assistant secretary, Atchison, Topeka and Santa Fe Railway Company, Strong correspondence, September–October 1881.

234 *"Ever since the first day . . .":* Ibid.

235 *"Mohave is better . . .":* Ibid., Strong correspondence March-July 1882.

235 *In late July, Strong fell sick:* Ibid.

236 *He confided to Wheeler:* Ibid., Strong correspondence July-October 1882.

236 *On November 11, Strong admitted:* Ibid., Strong correspondence November 1882-October 1883.

CHAPTER 25: THE END OF THE LINE

238 *When the owners refused:* Blair, Leadville: Colorado's Magic City, 164–170.

238 *He ended the debacle:* Cook, Hands Up!, 349.

238 *The dispute also took:* Blair, Leadville: Colorado's Magic City, 170.

238 *Although the* Rocky Mountain News: Burke, The Legend of Baby Doe, 106.

239 *As a wedding present:* Ibid., 114–116.

240 *After Tabor died in 1899:* Ibid., 161–224.

240 *To one newspaper correspondent:* Athearne, Rebel of the Rockies, 101.

240 *For one stretch, the mountainside:* Ibid., 101.

240 *In his rush, he'd sent:* Ibid., 109.

241 *Palmer tried furiously:* Ibid., 110–111.

241 *To save himself:* Ibid., 115.

241 *Annoyed to see Palmer's combined:* Ibid., 125.

244 *Happily for the RG investors:* Ibid., 133.

244 *Refusing to concede:* Ibid., 135.

244 *The Rio Grande board replaced:* Fisher, *A Builder of the West,* 266.

245 *By then, the Rio Grande finances:* Athearne, *Rebel of the Rockies,* 135.

245 *The lessee turned on his lessor:* Ibid.,140.

245 *When the court sided:* Ibid., 145.

245 *When Lovejoy put in scab:* Ibid., 151–152.

246 *It was bought for pennies:* Ibid., 153.

CHAPTER 26: IGHTHAM MOTE

248 *He professed to be full:* Ibid., 394.

249 *From there, she moved:* Black, *Queen of Glen Eyrie,* 97.

249 *The interior surrounded:* Ibid., 100.

250 *She made it a cultural retreat:* Gehling, *Queen & her General,* loc. 423, Kindle.

251 *When Queen reported to the General:* Black, *The Queen of Glen Eyrie,* 88–90.

251 *"My big darling . . .":* Ibid., 29–30.

251 *In 1887, Queen told Palmer:* Black, *The Queen of Glen Eyrie,* 104.

253 *In 1891 she confided:* Lucey, *Sargent's Women: Four Lives Behind the Canvas,* 37.

253 *Something of an artist himself:* Ibid., 60.

254 *"Do wear it . . .":* Ibid., 55.

254 *In a late volume:* Myers, *The Root and the Flower,* 201.

255 *His own finances were well established:* Fisher, *A Builder of the West,* 303–304.

255 *He devoted himself to local philanthropy:* Ibid., 306–307.

255 *"It seems to be . . .":* Lucey, *Sargent's Women: Four Lives Behind the Canvas,* 64.

256 *A photograph shows Palmer:* Fisher, *A Builder of the West,* 312–317.

256 *One March day in 1909:* Fisher, *A Builder of the West,* 314–318.

256 *A year later, his children:* In presenting a somewhat dark view of the Palmer marriage, I recognize I am flying in the face of the far more romantic accounts of a half-dozen books that make up the majority view. At the risk of sounding cynical, I simply have not been able to bring myself to see it that way, as I hope my account will justify. I say this with some regret, since I would have loved to go with the conventional wisdom. As Hemingway might say, it is prettier to think so, especially when you consider the effort that Palmer, at least, put into making the marriage a thing of beauty. The truth is that it was merely poignant.

CHAPTER 27: CALIFORNIA FOR A DOLLAR

257 *His first letter from the period:* Files of the assistant secretary, Atchison, Topeka and Santa Fe Railway Company, Strong correspondence, November 1881–October 1883.

257 *If the route had not come:* Bryant, *Atchison, Topeka and Santa Fe Railway,* 93.

258 *"Two more different men . . .":* Rayner, *The Associates,* loc. 1948, Kindle.

258 *After his long-suffering wife:* Ibid., loc. 1849–1853.

258 *As he strained to extend:* Ibid., loc. 1859–1866.

258 *His Santa Fe was still stuck:* Bryant, *Atchison, Topeka and Santa Fe Railway,* 102.

259 *As a condition of the sale:* Ibid., 102.

260 *On May 31, 1887, the first:* Ibid., 103.

261 *When Huntington charged $125:* Borneman, *Rival Rails,* 264.

261 *By 1890, the L.A. population:* Guinn and Beck, A *History of California and Extended History of Los Angeles,* 286–291.

CHAPTER 28: BOOM!

262 *"What do you want . . .":* "One Hundred Years of American History," *The American Monthly Magazine,* 163.

262 *It was, said one:* Belich, *Replenishing the Earth,* 339.

263 *"The sky is of a clear, bright . . .":* P. C. Remondino, MD, *The Mediterranean Shores of America, Southern California,* 112.

263 *Given his employers interests:* Nordhoff, *California for Health, Pleasure, and Residence,* 112–113.

264 *He deemed the region:* Truman, *Semi-tropical California,* 80–188.

264 *It was left to a later journalist:* Kevin Starr, *California and the American Dream,* loc. 968, Kindle. I need to add that no historical account of the development of California is possible without reference to Starr, who made a brilliant career of understanding the Golden State over nine books.

264 *Warner argued our Italy:* Warner, *Our Italy,* 11, https://tile.loc.gov/storage-services//service/gdc/calbk/168.pdf.

264 *In just the first six months:* Ibid, 4.

264 *A typical lot on Seventh Street:* Dumke, *The Boom of the Eighties in Southern California,* 44.

266 *There was supposed to be:* Ibid., 183.

266 *"A medley of colors was blazing . . .":* Van Dyke, *Millionaires of a Day,* 68.

267 *They appeared "like scenes . . .":* McWilliams, *Southern California,* 120.

268 *The first grand hotel:* Scheide, *Pasadena,* 70–71.

268 *One lot that had sold:* Ibid., 62.

268 *Five trains a day ran:* Dumke, "The Real Estate Boom of 1887 in Southern California", 431–432.

268 *And the city dressed up:* Scheide, *Pasadena,* 121, 143.

269 *No one found exactly:* Farmer, *Trees in Paradise,* 234–235. I'd be remiss if I failed to acknowledge a heavy debt to Jared Farmer for his *Trees in Paradise,* which is by far the best account of California's extraordinary—and defining—horticultural history.

269 *He thought the couple:* Ibid., 235.

270 *After the Tibbets Navel:* Ibid., 237.

270 *Orange-growing towns sprang:* Holmes, *History of Riverside County, California,* 56–57.

270 *When the town of Riverside:* Starr, *Inventing the Dream,* 146.

270 *By 1890, Riverside alone shipped:* Warner, *Our Italy,* 120.

270 *Soon came lemons, peaches, plums:* Farmer, *Trees in Paradise*, 350.

271 *Originally to be found:* Ibid., 336–350.

271 *By the time of Chicago's famous:* Ibid., 336.

271 *When the Santa Fe dropped off:* Scheid, *Pasadena*, 71.

CHAPTER 29: HOME

272 *As the population of the city:* Lothrop, "The Boom of the '80s Revisited," 269.

272 *That year, the Santa Fe's total:* Glischinski, *Santa Fe Railway*, 23.

273 *It started with the Santa Fe's Harvey Houses:* Bryant, *Atchison, Topeka and Santa Fe Railway*, 122.

276 *Once just a local offering:* "Building an American Icon," https://www.anheuser busch.com/about/heritage.html.

277 *"The history of western railroad . . .":* Bradley, *The Story of the Santa Fe*, 187–190.

278 *Santa Fe stock plunged:* Bryant, *Atchison, Topeka and Santa Fe Railway*, 150.

278 *For the first time, industry analysts:* Ibid., 150.

278 *Refusing to be reduced:* Bryant, *Atchison, Topeka and Santa Fe Railway*, 151.

278 *The official history of Strong's native:* Pack, "Rock County, Wisconsin," http://genealogytrails.com/wis/rock/history1908_pg3.html.

EPILOGUE: THE FRONTIER THESIS

281 *They named it Hollywood:* Starr, *Inventing the Dream*, loc. 6246–6271, Kindle.

281 *Horsley was in the nascent movie business:* "David Horsley," https://hollywood forever.com/story/david-horsley/.

282 *The Universal Film Company:* Ibid.

283 *It indicated that the American population:* "The significance of the frontier in American history," http://nationalhumanitiescenter.org/pds/gilded/empire/text1/turner.pdf.

283 *This became known:* Turner, "The Significance of the Frontier in American History," accessed at http://xroads.virginia.edu/~Hyper/TURNER/.

284 *This is part of what:* Mondi, " 'Connected and Unified?': A More Critical Look at Frederick Jackson Turner's America," 31.

284 *The original settlers, mostly from Europe:* Binding, "Huck Revisited," https://www.independent.co.uk/arts-entertainment/huck-revisited-1176103.html.

285 *After being taken over by Amtrak:* Simon, "The Short Troubled Life of Penn Central," http://passengertrainjournal.com/short-troubled-life-penn-central-passenger-trains/.

289 *In the course of laying:* Bradley, *The Story of the Santa Fe*, 331–371.

Bibliography

Abramo, Marisa and Mary Allely. "A Guide to the Kimball Family Collection." 1997. http://www.nationalcityca.gov/home/showdocument?id=145. Accessed online on July 25, 2020.

Adams, Charles Francis, Jr. *Railroads: Their Origins and Problems*. Kindle ed. Big Byte Books, 2016.

Alden, Henry Mills. "Air Towns and Their Inhabitants." *Harpers New Monthly Magazine*, vol. 51 (January 1875).

Ambrose, Stephen E. *Nothing Like it in the World: The Men Who Built the Transcontinental Railroad 1863–1869*. New York: Simon & Schuster, 2000.

"A Mountain Holiday." *McBride's Magazine*, vol. 21 (1878).

Anderson, George L. *General William J. Palmer*. Colorado Springs: Colorado College Publication, 1936.

Atchison, Topeka, and Santa Fe Railway Company Records. "William B. Strong Letters 1881–1884" and "Raton Pass Letters." Box 960. Kansas State Historical Society. Topeka, Kansas. Accessed in March 2018.

"Atchison, Topeka, and Santa Fe Railway Co. Land Grant Records." https://www.kshs.org/p/atchison-topeka-santa-fe-railroad-co-land-records/19951. Accessed online on July 25, 2020.

Athearn, Robert G. *Rebel of the Rockies: A History of the Denver and Rio Grande Western Railroad*. New Haven. Yale University Press, 1962.

"Back Bay Houses." https://backbayhouses.org/79-marlborough/. Accessed online on July 25, 2020.

Bain, David Haward. *Empire Express: Building the First Transcontinental Railroad*. New York: Penguin, 1999.

Bancroft, Caroline. *Silver Queen: The Fabulous Story of Baby Doe Tabor*. Wildside Press, 1950.

Bancroft, Hubert Howe. *History of Mexico, 1883–88*. San Francisco: The History Company Publishers, 1888.

Barker, Bill, and Jackie Lewin. *Denver! An Insider's Look at the High, Wide, and Handsome City*. New York: Doubleday & Company, Inc., 1972.

Barnes, Lela, ed. "Letters of Cyrus Kurtz Holliday," 1854–1869, Kansas Historical Quarterly, August 1937. https://www.kshs.org/p/letters-of-cyrus-kurtz-holliday-1854-1859/12717. Accessed online on July 25, 2020.

Barth, Gunther. *Instant Cities: Urbanization and the Rise of San Francisco and Denver.* New York: Oxford University Press, 1975.

Bartkey, Ian R. *Selling the True Time: Nineteenth-Century Timekeeping in America.* Stanford: Stanford University Press, 2000.

Beebe, Lucius and Charles Clegg. *Rio Grande: Mainline of the Rockies.* Berkeley: Howell-North Books, 1962.

Belich, James. *Replenishing the Earth: The Settler Revolution and the Rise of the Anglo-World, 1783–1939.* New York: Oxford University Press, 2009.

Bell, William Abraham. *New Tracks in North America.* London: Chapman and Hall, 1869.

Bellesiles, Michael A. *1877: America's Year of Living Violently.* Kindle ed. New York: The New Press, 2010.

Berkman, Pamela, ed. *The History of the Atchison, Topeka, & Santa Fe.* Hong Kong: Brompton Books Corp., 1988.

Binding, Paul. "Huck Revisited." October 4, 1998. https://www.independent.co.uk/arts-entertainment/huck-revisited-1176103.html. Accessed online on July 25, 2020.

Black, Celeste. *Queen of Glen Eyrie.* Colorado Springs: Nav Press, 2008.

Blair, Edward. *Leadville: Colorado's Magic City.* Boulder: Fred Pruett Books, 1980.

Blevins, Tim and Dennis Daily, Chris Nicholl, Calvin P. Otto, Katherine Scott Sturdevant, eds. *Legends, Labors & Loves: William Jackson Palmer 1836–1909.* Colorado Springs: Pikes Peak Library District, 2009.

Borneman, Walter R. *Iron Horses: America's Race to Bring the Railroads West.* New York: Back Bay Books, 2010.

Borneman, Walter R. *Marshall Pass: Denver & Rio Grande Gateway to the Gunnison Country.* Colorado Springs: Century One Press, 1980.

Borneman, Walter R. *Rival Rails: The Race to Build America's Greatest Transcontinental Railroad.* New York: Random House, 2010.

Bowers, John. *Chikamauga and Chattanooga: The Battles That Doomed the Confederacy.* New York: Avon Books, 1994.

Boyd, Conseulo. "Twenty Years to Nogales: The Building of the Guaymas-Nogales Railroad." *The Journal of Arizona History,* vol. 22, no. 3 (Autumn 1981).

Bradley, Glenn. *The Story of the Santa Fe.* Palmdale: Omni Publications, 1995.

Brands, H.W. *American Collossus: The Triumph of Capitalism 1865–1900.* Kindle ed. New York: Doubleday, 2010.

Brandt, Walther Immanuel. "The Railway Invasion of Mexico." A thesis submitted for Master of Arts to the University of Wisconsin, 1917.

Briggeman, Kim. "Montana History Almanac: Trader killed for alleged insult of Piegan Indians." August 13, 2011. https://missoulian.com/lifestyles/territory/montana-history-almanac-trader-killed-for-alleged-insult-to-piegan/article_48c9b0a0-c51d-11e0-ae35-001cc4c002e0.html. Accessed online on July 25, 2020.

Brown, Dee. *Hear that Lonesome Whistle Blow: Railroads in the West*. New York: Holt, 1977. "Bryan's Cross of Gold Speech." http://historymatters.gmu.edu/d/5354/. Accessed online on July 25, 2020.

Bryant, Jr. Keith L. *History of the Atchison, Topeka and Santa Fe Railway*. Lincoln: University of Nebraska Press, 1974.

"Building an American Icon." https://www.anheuser-busch.com/about/heritage.html. Accessed online on July 25, 2020.

Burke, John. *The Legend of Baby Doe: The Life and Times of the Silver Queen of the West*. Lincoln: University of Nebraska Press, 1974.

Carubia, Josephine, and Lorraine Dowler, Bonj Szczygiel, eds. *Gender and Landscape: Renegotiating the Moral Landscape*. New York: Routledge, 2005.

"CB&Q Builds to Denver." https://history.nebraska.gov/sites/history.nebraska.gov/files/doc/publications/NH1959CBQtoDenver.pdf. Accessed online on July 25, 2020.

Chandler, Alfred D. *The Visible Hand: The Managerial Revolution in American Business*. Cambridge: The Belknap Press, 1977.

"Chronology of Railroading in America." https://www.aar.org/data/chronology-rail roading-america/. Accessed online on July 25, 2020.

Clavin, Tom. *Dodge City: Wyatt Earp, Bat Masterson, and the Wickedest Town in the American West*. Kindle ed. New York: St. Martin's Press, 2017.

Cleaveland, Agnes Morley. *No Life for a Lady*. Lincoln: University of Nebraska Press, 1941.

Cleaveland, Norman, and George Fitzpatrick. *The Morleys—Young Upstarts on the Southwest Frontier*. Albuquerque: Calvin Horn Publisher, Inc., 1971.

Clothier, Isaac and William Palmer. *General William J. Palmer: Letters, 1853–1868 (1906)*. Philadelphia: Kessinger Publishing, 1906.

Colorado Chieftain excerpted from Volume 7, Number 1703, November 8, 1877. https://www.coloradohistoricnewspapers.org/cgi-bin/colorado?a=d&d=CFT18771 108-01.2.18&e=-en-20img-txIN%7ctxCO%7ctxTA. Accessed online on July 25, 2020.

Colton, Walter. *The Land of Gold; Or; Three Years in California*. New York: D.W. Evans & Co, 1860.

Committee on Commerce United States Senate. "Study of Essential Railroad Service." July 24, 25, and 29, 1968. Serial No. 91–81.

Conrad, Howard L. *"Uncle Dick" Wootton*. Chicago: W. E. Dibble & Co., 1890.

Cook, David. *Hands Up: Or, Twenty Years of Detective Life on the Mountains and on the Plains*. Kindle ed. Project Gutenberg, 2013.

Coolidge, T. Jefferson. *The Autobiography of T. Jefferson Coolidge*. Massachusetts Historical Society, 1923.

Cronon, William. *Nature's Metropolis: Chicago and the Great West*. New York: W. W. Norton & Company, 1991.

"David Horsley." https://hollywoodforever.com/story/david-horsley/. Accessed online on July 25, 2020.

Davis, E. O. *The First Five Years of the Railroad Era in Colorado.* Sage Books, 1948.

DeArmont, Robert. *Bat Masterson: The Man and the Legend.* Kindle ed. Norman: University of Oklahoma Press, 1988.

D'Emilio, Sandra, and Susan Campbell. *Visions & Visionaries: The Art & Artists of the Santa Fe Railway.* Salt Lake City: Peregrine Smith Books, 1991.

Deverell, William. *Railroad Crossing: Californians and the Railroad 1850–1910.* Berkeley: University of California Press, 1994.

Dodge, Grenville M. *How We Built the Union Pacific Railroad.* Big Byte Books, 2014.

Dorset, Phyllis Flanders. *The New Eldorado: The Story of Colorado's Gold and Silver Rushes.* Golden: Fulcrum Publishing, 1970.

Ducker, James H. *Men of the Steel Rails: Works on the Atchison, Topeka & Santa Fe Railroad, 1869–1900.* Lincoln: University of Nebraska Press, 1983.

Durian, Hal. *True Stories of Riverside and the Inland Empire.* Charleston: The History Press, 2013.

Dumke, Glenn S. *The Boom of the Eighties in Southern California.* San Marino: Huntington Library, 1944.

Dumke, Glenn S. "The Real Estate Boom of 1887 in Southern California." *Pacific Historical Review,* vol. 11, no. 4 (December 1942).

Ellmann, Richard. *Oscar Wilde.* New York: Vintage Books, 1988.

Emerson, Ralph Waldo. *Journals of Ralph Waldo Emerson.* Cambridge: The Belknap Press, 1964. "Express Train Crosses the Nation in 83 Hours." https://www.history .com/this-day-in-history/express-train-crosses-the-nation-in-83-hours. Accessed online on July 25, 2020.

Farmer, Jared. *Trees in Paradise: A California History.* Kindle ed. New York: W. W. Norton & Company, 2013.

Ferrell, Mallory Hope. *Denver & Rio Grande: The Early Years.* Bucklin: Whiteriver Productions, 2018.

Fishel, Edwin C. *The Secret War for the Union: The Untold Story of Military Intelligence in the Civil War.* New York: Houghton Mifflin Company, 1996.

Fisher, John S. *A Builder of the West: The Life of General William Jackson Palmer.* Caldwell, The Caxton Printers, Ltd. 1939.

Fogarty, Robert S. *All Things New: American Communes and Utopian Movements, 1860–1914.* Lanham: Lexington Books, 2003.

Fogelson, Robert M. *America's Armories: Architecture, Society, and Public Order.* Cambridge: Harvard University Press, 1989.

Fogelson, Robert M. *The Fragmented Metropolis: Los Angeles, 1850–1930.* Berkeley: University California Press, 1967.

Fried, Stephen. *Appetite for America.* Kindle ed. New York: Bantam Books, 2011.

Furman, Evelyn E. Livingston. *Silver Dollar Tabor: The Leaf in the Storm.* Englewood: Quality Press, 1982.

Gandy, Lewis Cass. *The Tabors: A Footnote of Western History.* New York: The Press of the Pioneers, Inc. 1934.

Gates, Paul Wallace. *Fifty Million Acres: Conflicts over Kansas Land Policy, 1854–1890*. Norman: University of Oklahoma Press, 1997.

Gehling, Richard and Mary Ann. *Queen & Her General*. Kindle ed. Amazon Services, LLC, 2010.

"General Time Convention." https://chicagology.com/transportation/timeconvention/. Accessed online on July 25, 2020.

Glass, Andrew. "President McKinley Signs Gold Standard Act, March 14, 1990." *Politico*. March 14, 2013. https://www.politico.com/story/2013/03/this-day-in-politics-088821. Accessed online on July 25, 2020.

Glischinski, Steve. *Santa Fe Railway*. Osceola: MBI Publishing Company, 1997.

Goetzmann, William. *Exploration and Empire: The Explorer and the Scientist in the Winning of the American West*. Austin: Monticello Editions, 1993.

Goodrich, Thomas. *War to the Knife: Bleeding Kansas: 1854–1861*. Lincoln: University of Nebraska Press, 1988.

Griswold, Don L. and Jean Harvey. *The Carbonate Camp Called Leadville*. Denver: The University of Denver Press, 1951.

Grodinsky, Julius. *Jay Gould: His Business Career, 1867–1892*. New York: Arno Press, 1981.

———. *Transcontinental Railway Strategy, 1869–1893: A Study of Businessmen*. Philadelphia: University of Pennsylvania Press, 1962.

Guinn, James Miller and Juergen Beck. *A History of California and an Extended History of Los Angeles*. Kindle ed. Loschberg: Jazzybee, 2015.

Harper, Jared V. *Santa Fe's Raton Pass*. Dallas: Kachina Press, 1983.

Hart, John Mason. *Empire and Revolution: The Americans in Mexico since the Civil War*. Berkeley: University of California Press, 2002.

———. *Revolutionary Mexico: The Coming and Process of the Mexican Revolution*. Berkeley: University of California Press, 1987.

Hayes, Derek. *Historical Atlas of California*. Berkeley: University of California Press, 2007.

Higgins, Charles A. *To California over the Santa Fe Trail*. Chicago: Passenger Department, 1914.

Hilton, George. 1 May 2006. "A History of Track Gauge." https://trn.trains.com/rail roads/abcs-of-railroading/2006/05/a-history-of-track-gauge. Accessed online on July 25, 2020.

Hirshson, Stanley P. *Grenville M. Dodge: Soldier, Politician, Railroad Pioneer*. Bloomington: Indiana University Press, 1967.

"Historic El Paso Population." https://txcip.org/tac/census/hist.php?FIPS=48141. Accessed online on July 25, 2020.

History.com editors. "The California Gold Rush." https://www.history.com/topics/west ward-expansion/gold-rush-of-1849. Accessed online on July 25, 2020.

Hittell, John. *A History of the City of San Francisco*. First Rate Publishers.

Hodges, Gladys A. "Bridges Across the Borderline: The Local Politics of Building the First International Rail Bridges in the Americas at the Two El Pasos, 1880–1883." *The Southwestern Historical Quarterly*, vol. 116, no. 1 (July 2020).

Holbrook, Stewart H. *The Golden Age of Railroads*. New York: Random House, 1960.

Holmes, Elmer Wallace. *History of Riverside County, California*. North Charleston: Createspace.

Hooper, S.K. *The Story of Manitou*. Denver & Rio Grande R.R., 1980.

Howard, Kathleen L. and Diana Pardue. *Inventing the Southwest: The Fred Harvey Company and Native American Art*. Flagstaff: Northland Publishing Company, 1996.

Hudson, John C. *Plains Country Towns*. Minneapolis: University of Minnesota Press, 1985.

————. "Towns of the Western Railroads." *Great Plains Quarterly*, Winter 1982.

Ingersoll, Ernest. *The Crest of the Continent*. Chicago: R. R. Donnelley & Sons Company, 1890.

Inman, Henry. *The Old Santa Fé Trail, The Story of a Great Highway*. Kindle ed. A public domain book.

Jackson, Robert Wendell. *Rails Across the Mississippi: A History of the St. Louis Bridge*. Urbana: University of Illinois Press, 2001.

Johnson, Arthur M. and Barry E. Supple. *Boston Capitalists and Western Railroads: A Study in the Nineteenth-Century Railroad Investment Process*. Cambridge: Harvard University Press, 1967.

Karsner, David. *Silver Dollar: The Story of the Tabors*. New York: Crown Publishers, 1966.

Kingsley, Charles. *South by West; or, Winter in the Rocky Mountains and Spring in Mexico*. London: W. Isbister & Co., 1874.

Klein, Maury. *The Life and Legend of Jay Gould*. Baltimore: The John Hopkins University Press,1986.

Knapp, Frank A. Jr. "Precursors of American Investment in Mexican Railroads." *Pacific Historical Review*, vol. 21, no. 1 (February 1952).

Latimer, Rosa Walston. *Harvey Houses of New Mexico: Historic Hospitality from Raton to Deming*. Charleston: The History Press, 2015.

Lavender, David. *The Great Persuader*. New York: Doubleday & Company, Inc., 1970.

Lavender, David. *The Southwest*. Albuquerque: University of New Mexico Press, 1980.

"Leadville, Colorado." https://westernmininghistory.com/towns/colorado/leadville/. Accessed online on July 25, 2020.

Leadville Daily/ Evening Chronicle, vol. 1, no. 33 (March 8, 1879). https://www .coloradohistoricnewspapers.org/?a=d&d=LEC18790308-01.1.1. Accessed online on July 25, 2020.

"Legends of America." https://www.legendsofamerica.com/we-clayallison/2/. Accessed online July 25, 2020.

Lewis, Oscar. *The Big Four*. New York: Alfred A. Knopf, 1938.

Llaso, Steve. "Denver & Rio Grande." https://www.steamlocomotive.com/locobase
.php?country=USA&wheel=2-6-0&railroad=drgw#11704. Accessed online on
July 25, 2020.

Lohse, Joyce B. *Baby Doe Tabor: Matchless Silver Queen*. Palmer Lake: Filter Press
Books, 2011.

———. *General William Palmer: Railroad Pioneer*. Palmer Lake: Filter Press Books,
2009.

Lotchin, Roger W. *San Francisco, 1846–1856: From Hamlet to City*. Urbana: University
of Illinois Press, 1997.

Lothrop, Gloria Ricci. "The Boom of the '80s Revisited." *Southern California Quarterly*,
vol. 75, nos. 3–4 (October 1993).

Lowry, Thomas, M.D. *The Story the Soldiers Wouldn't Tell: Sex in the Civil War*. Kindle
ed. Mechanicsburg: Stackpole Books, 1994.

Lucey, Donna M. *Sargent's Women: Four Lives Behind the Canvas*. New York: W. W.
Norton & Company, 2017.

Lyman, Edward L. "Outmaneuvering the Octopus: Atchison, Topeka, and Santa Fe."
California History, vol. 67, no. 2 (June 1988).

Marshall, James. *Santa Fe: The Railroad That Built an Empire*. New York: Random
House, 1945.

Matthews, Michael. *The Civilizing Machine: A Cultural History of Mexican Railroads,
1876–1910*. Lincoln: University of Nebraska Press, 2013.

May, Stephen J. *A Kingdom of Their Own: The Story of the Palmers of Glen Eyrie*.
Johnson Books, 2016.

McWilliams, Carey. *California: The Great Exception*. Kindle ed. Berkeley: University
of California Press, 1949.

McWilliams, Carey. *Southern California: An Island on the Land*. Salt Lake City:
Peregrine Smith Books, 1973.

Miller, Darlis A. *Open Range: The Life of Agnes Morley Cleaveland*. Kindle ed. Norman:
University of Oklahoma Press, 2010.

Moehring, Eugene P. *Urbanism and Empire in the Far West, 1840–1890*. Reno:
University of Nevada Press, 2004.

Mondi, Megan. "'Connected and Unified?': A More Critical Look at Frederick Jackson
Turner's America." *Constructing the Past*, vol. 7, no. 1 (2006).

Montoya, Maria E. *Translating Property: The Maxwell Land Grant and the Conflict over
Land in the American West, 1840–1900*. Lawrence: University Press of Kansas,
2002.

Morin, Karen M. *Frontiers of Femininity: A New Historical Geography of the Nineteenth-
Century American West*. Syracuse: Syracuse University Press, 2008.

Morris, Charles R. *The Tycoons: How Andrew Carnegie, John D. Rockefeller, Jay Gould,
and J.P. Morgan Invented the American Supereconomy*. New York: Henry Holt &
Company, 2005.

Murphy, Lawrence. *Lucien Bonaparte Maxwell: Napoleon of the Southwest*. Norman: University Oklahoma Press, 1983.

"Mussel Slough Incident." https://www.encyclopedia.com/history/dictionaries-the saruses-pictures-and-press-releases/mussel-slough-incident. Accessed online on July 25, 2020.

Myers, Leopold Hamilton. *The Root and the Flower*. New York: New York Review of Books, 1985.

"New Mexico's Lincoln County War." https://www.legendsofamerica.com/nm -lincolncountywar/. Accessed online on July 25, 2020.

Noble, Andrew, ed. *From the Clyde to California: Robert Louis Stevenson's Emigrant Journey*. Aberdeen: Aberdeen University Press, 1985.

Nordhoff, Charles. *California for Health, Pleasure and Residence*. New York: Harper & Brothers, Publishers, 1873.

Norris, Frank. *The Octopus: A California Story*. Kindle ed. Digireads.com Publishing, 2009. Office of Corporate Communications. "U.S. Mint History: The Crime of '1873,'" March 22, 2017. https://www.usmint.gov/news/inside-the-mint/mint -history-crime-of-1873. Accessed online on July 25, 2020.

"One Hundred Years of American History." *The American Monthly Magazine*. Vol. XIV, no. 2, February 1899.

Orsi, Richard J. *Sunset Limited: The Southern Pacific Railroad and the Development of the American West 1850–1930*. Berkeley: University of California Press, 2005.

Osterwald, Doris B. *High Line to Leadville: A Mile by Mile Guide for the Leadville, Colorado & Southern Railroad*. Conifer: Western Guideways, Ltd., 1991.

———. *Rails thru the Gorge: A Mile by Mile Guide for the Royal Gorge Route*. Lakewood: Western Guideways, Ltd., 2003.

Pack, Andrea Stawski. "Rock County, Wisconsin History." http://genealogytrails.com /wis/rock/history1908_pg3.html. Accessed online on July 25, 2020.

Palmer, William Jackson. *Report of Surveys Across the Continent in 1867–'68, on the Thirty-Fifth and Thirty-Second Parallels, For a Route Extending the Kansas Pacific Railway to the Pacific Ocean at San Francisco and San Diego*. Philadelphia: W.B. Selheimer, Printer, 1869.

Paul, Rodman Wilson. *Mining Frontiers of the Far West, 1848–1880*. Albuquerque: University of New Mexico Press, 1963.

Perry, Thomas D. *"If Thee Must Fight, Fight Well."* Ararat: Laurel Hill Publishing, 2013.

Pfeiffer, David A. "Bridging the Mississippi: The Railroads and Steamboats Clash at the Rock Island Bridge." *Prologue Magazine*, vol. 36, no. 2 (Summer 2004). https://www.archives.gov/publications/prologue/2004/summer/bridge.html Accessed online on July 25, 2020.

Pierce, Jason. "Our Climate and Soil Is Completely Adapted to Their Customs" from *Making the White Man's West*. https://www.jstor.org/stable/pdf/j.ctt19jcg63.12 .pdf. Accessed online July 25, 2020.

Pisani, Bob. "This Single Paged Document Started the New York Stock Exchange 225 Years ago." May 17, 2017. https://www.cnbc.com/2017/05/17/this-single -paged-document-started-the-new-york-stock-exchange-225-years -ago.html. Accessed online on July 25, 2020.

Pletcher, David M. "General William S. Rosecrans and the Mexican Transcontinental Railroad Project." *The Mississippi Valley Historical Review*, vol. 38, no. 4 (March 1952).

———. "Mexico Opens the Door to American Capital, 1877–1880." *The Americas*, vol. 16, no. 1 (July 1959).

———. *Rails, Mines, and Progress: Seven American Promoters in Mexico, 1867–1911*. Port Washington: Kennikat Press, 1958.

Pretti, Roger. *Mining, Mayhem and Other Carbonate Excitements: Tales from a Silver Camp Called Leadville*. Vail: Vail Daily, 2003.

Prerau, David. *Seize the Daylight: The Curious and Contentious Story of Daylight Saving Time*. New York: Thunder's Mouth Press, 2005.

Quiett, Glenn Chesney. *They Built the West: An Epic of Rails and Cities*. New York: D. Appleton-Century Company, 1934.

"Railroad Beginnings in California." https://www.railswest.com/history/california beginnings.html. Accessed online on July 25, 2020.

"Railroads, Federal Land Grants, To . . ." https://www.encyclopedia.com/history /encyclopedias-almanacs-transcripts-and-maps/railroads-federal-land-grants -issue. Accessed online on July 25, 2020.

Railway Review, vol. 20, no. 1–52 (August 7, 1880).

Rasmussen, Stephen. *The Rio Grande's La Veta Pass Route: Gateway to the San Luis Valley*. Burlington: Evergreen Press, 2000.

Rayner, Richard. *The Associates: Four Capitalists Who Created California*. Kindle ed. New York: W. W. Norton & Company, 2008.

Remondino, Peter. *The Mediterranean Shores of America*. Philadelphia: The F. A. Davis Co., Publishers, 1892.

Renehan, Edward. *Dark Genius of Wall Street: The Misunderstood Life of Jay Gould, King of the Robber Barons*. Kindle ed. New York: Basic Books, 2005.

Reps, John W. *Cities of the American West: A History of Frontier Urban Planning*. Princeton: Princeton University Press, 1979.

———. *The Forgotten Frontier: Urban Planning in the American West Before 1980*. Columbia: University of Missouri Press, 1981.

Richter, Amy G. *Home on the Rails: Women, the Railroad, and the Rise of Public Domesticity*. Chapel Hill: University of North Carolina Press, 2005.

Riegel, Robert Edgar. *The Story of the Western Railroads: From 1852 through the Reign of the Giants*. Lincoln: University of Nebraska Press, 1926.

"Rise of Industrial America." http://www.loc.gov/teachers/classroommaterials/presen tationsandactivities/presentations/timeline/riseind/railroad/. Accessed online on July 25, 2020.

Scheid, Ann. *Pasadena: Crown of the Valley*. (Albany, NY: Windsor Publications, 1986.)

Schivelbusch, Wolfgang. *The Railway Journey: The Industrialization of Time and Space in the Nineteenth Century*. Kindle ed. Oakland: University of California Press, 1977.

Schwantes, Carlos A. *The Pacific Northwest*. Lincoln: University of Nebraska, 1996.

Schwantes, Carlos A. and James P. Ronda. *The West the Railroads Made*. Seattle: University of Washington Press, 2008.

Simon, El. "The Short, Troubled Life of Penn Central Passenger Trains," February 2018. *Passenger Train Journal*. http://passengertrainjournal.com/short-troubled -life-penn-central-passenger-trains/. Accessed online on July 25, 2020.

Simonin, Louis L. *The Rocky Mountain West in 1867*. Lincoln: University of Nebraska Press, 1966.

Smith, Duane A. *Horace Tabor: His Life and the Legend*. Niwot: University Press of Colorado, 1989.

Smythe, William Ellsworth. *History of San Diego 1542–1908*. Kindle ed. Loschberg: Jazzybee.

Solomon, Brian. *Railroads of California: The Complete Guide to Historic Trains and Railway Sites*. Minneapolis: Voyageur Press, 2009.

Sprague, Marshall. *Newport in the Rockies*. Chicago: Sage/Swallow Press Books, 1981.

———. *The Great Gates: The Story of the Rocky Mountain Passes*. Lincoln: University of Nebraska Press, 1964.

Stanislawski, Dan. "The Origin and Spread of the Grid-Pattern Town." *Geographical Review*, vol. 36, no. 1 (January 1946).

Starr, Kevin. *Americans and the California Dream, 1850–1915*. Kindle ed. New York: Oxford University Press, 1973.

———. *California: A History*. New York: A Modern Library Chronicles Book, 2007.

———. *Inventing the Dream: California Through the Progressive Era*. Kindle ed. New York: Oxford University Press, 1985.

Stegner, Wallace. *Beyond the Hundredth Meridian: John Wesley Powell and the Second Opening of the West*. New York: Penguin Books, 1953.

Stiles, T. J. *The First Tycoon: The Epic Life of Cornelius Vanderbilt*. New York: Vintage Books, 2009.

Stilgoe, John R. *Metropolitan Corridor: Railroads and the American Scene*. New Haven: Yale University Press, 1983.

Temple, Jude Nolte. *Baby Doe Tabor: The Madwoman in the Cabin*. Norman: University of Oklahoma Press, 2007.

"The 1870 Census." https://www.census.gov/library/publications/1872/dec/1870a.html. Accessed online on July 25, 2020.

"The Annexation of Texas." https://history.state.gov/milestones/1830-1860/texas -annexation. Accessed online on July 25, 2020.

"The Price of Gold." https://www.measuringworth.com/datasets/gold/result.php?gold silver=on&year_source=1687&year_result=2020. Accessed online on July 25, 2020.

"The Significance of the Frontier in American History." http://nationalhumanities center.org/pds/gilded/empire/text1/turner.pdf. Accessed online on July 25, 2020.

Todd, Nancy L. *New York's Historic Armories: An Illustrated History*. Albany: State University of New York, 2006.

Train, George Francis. *My Life in Many States and in Foreign Lands, Dictated in My Seventy-Fourth Year*. New York: D. Appleton and Company, 1902. http://www .gutenberg.org/ebooks/38265. Accessed online on July 25, 2020.

Treadway, William E. *Cyrus K. Holliday: A Documentary Biography*. Topeka: Kansas State Historical Society.

"A Tribute to Sheriff Cook." https://www.arapahoegov.com/DocumentCenter/View /8440/A-tribute-to-Sheriff-Cook. Accessed online on July 25, 2020.

Truman, Benjamin. *Semi-Tropical California*. San Francisco: A. L. Bancroft & Company, Publishers, 1874.

Turner, Frederick Jackson. "The Significance of the Frontier in American History." http://xroads.virginia.edu/~Hyper/TURNER/. Accessed online on July 25, 2020.

Twain, Mark and Charles Dudley Warner. *The Gilded Age: A Tale of Today*. New York: Penguin Books, 2001.

Unger, Irwin. *The Greenback Era: A Social and Political History of American Finance, 1865–1879*. Princeton: Princeton University Press, 1964.

Van Dyke, Theodore. *Millionaires of a Day*. New York: Fords, Howard & Hulbert, 1890.

Warman, Cy. *The Story of the Railroad*. Kindle ed. New York: D. Appleton and Company, 1898.

Warner, Charles Dudley. *Our Italy*. https://tile.loc.gov/storage-services//service/gdc /calbk/168.pdf. Accessed online on July 25, 2020.

Waters, L. L. and Deane Malott. *Steel Trails to Santa Fe*. Literary Licensing, 2012.

Wedel, David C. *The Story of Alexanderwohl*. Goessel: Goessel Centennial Committee, 1999.

Weinstein, Allen. *Prelude to Populism: Origins of the Silver Issue, 1867–1878*. New Haven: Yale University Press, 1970.

Wellington, Arthur Mellen. *The Economic Theory of the Location of Railways*. New York: Wiley, 1887.

Whitaker, Milo Lee. *Pathbreakers and Pioneers of the Pueblo Region*. The Franklin Press Company, 1917.

White, Richard. *Railroaded: The Transcontinentals and the Making of Modern America*. New York: W. W. Norton & Company, Inc., 2011.

Wilcox, Rhoda Davis. *The Man on the Iron Horse*. Manitou Springs: Martin Associates, 1959.

William Jackson Palmer Records. "Correspondence 1878–1879." Box 311. Denver Public Library, Denver, Colorado. Accessed in September 2017.

Wilson, O. Meredith. *The Denver and Rio Grande Project, 1870–1901*. Salt Lake City: Howe Brothers, 1982.

Withers, Charles W. J. *Zero Degrees: Geographies of the Prime Meridian*. Cambridge: Harvard University Press, 2017.

Wolcott, Frances M. *Heritage of Years: Kaleidoscopic Memories*. New York: Minton, Balch & Company, 1932.

Wolmar, Christian. *The Great Railroad Revolution: The History of Trains in America*. New York: Public Affairs, 2012.

Wood, Dorothy. *Long Eye & the Iron Horse: A Biography of Grenville Dodge and the Union Pacific Railroad*. New York: Criterion Books, 1966.

Wrobel, David M. *Promised Lands: Promotion, Memory, and the Creation of the American West*. Lawrence: University Press of Kansas, 2002.

Wyatt Earp and Bat Masterson. Kindle ed. Charles Rivers Editors, 2012.

"Wyatt Earp dropped from Wichita Police Force." https://www.history.com/this-day-in -history/wyatt-earp-dropped-from-wichita-police-force. Accessed online on July 25, 2020.

Illustration Credits

Endpaper (#1)   American Bank Note Company & Atchison, Topeka and Santa
                 Fe Railroad Company. Map of the Atchison, Topeka and the
                 Santa Fe railroad system, 1899, G3701.P3 1899.A4, Library of
                 Congress Geography and Map Division, Washington, DC.
Endpaper (#2)   S. W. Eccles & Denver and Rio Grande Railway Company.
                 Map of the Denver & Rio Grande Railway, c1881, G4311.P3
                 1881.E25, Library of Congress Geography and Map Division,
                 Washington, DC.

Page:
    6   Carl Mathews Collection, Pikes Peak Library District, 005–375.
    8   William J. Palmer Family Photograph Collection, Item 106, PP88-42, Special
        Collections, Tutt Library, Colorado College.
   16   Kansas State Historical Society.
22–23   William Henry Jackson, Denver Public Library, Special Collections.
   26   Friends of the Cumbres & Toltec Scenic Railroad, Richard L. Dorman
        Collection, AMP01–036.
   36   Andrew Joseph Russell, Golden Spike Ceremony, Promontory Summit, Utah.
   40   Cartoon from Frank Leslie's illustrated newspaper, June 5, 1869, LC-USZ62-
        124424, Library of Congress Prints and Photographs Division, Washington, DC.
   42   Photograph by Myron Wood, © Pikes Peak Library District, 002-3290.
   49   Rhoda Wilcox Collection, Pikes Peak Library District, 214–11469.
   69   De Agostini Editorial / DEA / Biblioteca Ambrosiana / Getty Images.
   75   Harold Elderkin, Jesus Silva, and Uncle Dick Wootton, East Las Vegas, 1885,
        Courtesy of the Palace of the Governors Photo Archives (NMHM/DCA),
        Neg. #013112.
   80   Kansas State Historical Society.
   88   Peter Newark American Pictures / Bridgeman Images.
   91   Photograph by William Bell, Margaretta M. Boas Photograph Collection,
        Pikes Peak Library District, 001-6025.
  100   Colorado Springs Pioneers Museum.
  103   Colorado Springs Pioneers Museum.
  107   Colorado Springs Pioneers Museum.
  109   New Mexico State University Library, Archives and Special Collections.

120   William Henry Jackson, The Royal Gorge, Grand Cañon of the Arkansas, circa 1880, The J. Paul Getty Museum, Los Angeles. Digital image courtesy of the Getty's Open Content Program.

137   History Colorado, Denver, Colorado.

140   History Colorado, Denver, Colorado.

143   History Colorado, Denver, Colorado.

160   The Granger Collection.

173   Frederick Burr Opper, Illustration from *Puck*, March 29, 1882, LC-DIG-ppmsca-28461, Library of Congress Prints and Photographs Division, Washington, DC.

183   Colorado Springs Pioneers Museum.

187   Kansas State Historical Society.

191   NAU.PH.568.7493: Northern Arizona University, Cline Library [Emery Kolb Collection].

194–95   Copyright © circa 1910, *Los Angeles Times*. Used with Permission.

201   E. S. Glover & A.L. Bancroft & Company, Birds-eye view of Los Angeles, California, 1877, G4364.L8A3 1877.G61, Library of Congress Geography and Map Division, Washington, DC.

205   Stephen W. Shaw, Portrait of Collis P. Huntington,1873, Crocker Art Museum, E. B. Crocker Collection.

207   Jimmy Swinnerton, C. P. Huntington as an Octopus,1896. Print. Courtesy of the California History Room, California State Library, Sacramento, California.

212   San Bernardino Historical and Pioneer Society/San Bernardino History and Railroad Museum.

224   Alpha Stock/ Alamy Stock Photo.

231   © CORBIS / Corbis via Getty Images.

239   AP Photo / Denver Post.

248   Colorado Springs Pioneers Museum.

250   © Louisa Creed, Courtesy of Ightham Mote.

252   John Singer Sargent, Portrait of Miss Elsie Palmer, or A Lady in White, Collection of the Colorado Springs Fine Arts Center at Colorado College, FA 1969.3.

256   Colorado Springs Pioneers Museum.

265   H. B Elliott, Elliott Pub. Co. & Southern California Land Co., Los Angeles, California, 1891, G4364.L8A3 1891.E6, Library of Congress Geography and Map Division, Washington, DC.

267   Rick Thomas Collection, *The South Pasadenan News*.

273   Charles F. Dowd, System of national time and its application, by means of hour and minute indexes, to the national railway time-table, 1870. Courtesy of The University Library, University of Illinois at Urbana-Champaign.

282   H. H. Wilcox and Company, Map of Hollywood, 442300, 1887, Huntington Library, San Marino, California.

Index

Note: A page number in italics refers to an illustration; the letter *n* following a page number refers to a footnote

About the Author

John Sedgwick is best known for his bestselling family memoir, *In My Blood*; his account of the Cherokee at the time of the Trail of Tears, *Blood Moon*; and *War of Two*, which is about the fatal rivalry between Alexander Hamilton and Aaron Burr. *War of Two* won the Society of Cincinnati Prize and was a finalist for the George Washington Prize, both given for the year's best book on the Founding Fathers. An earlier work, *The Peaceable Kingdom*, was the basis of a television series on CBS. He has published fifteen books altogether, including two novels, three works of literary nonfiction, and six collaborations. A longtime magazine writer, Sedgwick has been a contributing editor for *GQ* and *Newsweek*, and published frequently in *Esquire*, *Vanity Fair*, and the *Atlantic*. He is a member of the storied Sedgwick clan, which first arrived in America in 1636 and includes such notables as the early speaker of the house of representatives Theodore Sedgwick and the sixties fashion icon Edie Sedgwick. He is married to Rana Foroohar, the author, CNN analyst, and *Financial Times* columnist, and lives with Rana and her two children in Brooklyn, New York.